中国水旱灾害防御公报

China Flood and Drought Disaster Prevention Bulletin

2022

中华人民共和国水利部

Ministry of Water Resources of the People's Republic of China

中国水利水电出版社
www.waterpub.com.cn
·北京·

图书在版编目（CIP）数据

中国水旱灾害防御公报. 2022 / 中华人民共和国水利部编著. -- 北京 : 中国水利水电出版社, 2023.10
ISBN 978-7-5226-1841-8

Ⅰ. ①中… Ⅱ. ①中… Ⅲ. ①水灾－灾害防治－公报－中国－2022②干旱－灾害防治－公报－中国－2022
Ⅳ. ①P426.616

中国国家版本馆CIP数据核字(2023)第192473号

责任编辑：徐丽娟

审图号：GS 京 (2023)1497 号

书　名	中国水旱灾害防御公报 2022 ZHONGGUO SHUIHAN ZAIHAI FANGYU GONGBAO 2022
作　者	中华人民共和国水利部
出版发行	中国水利水电出版社 （北京市海淀区玉渊潭南路 1 号 D 座　100038） 网址：www.waterpub.com.cn E-mail：sales@mwr.gov.cn 电话：（010）68545888（营销中心）
经　售	北京科水图书销售有限公司 电话：（010）68545874、63202643 全国各地新华书店和相关出版物销售网点
排　版	北京水精灵教育科技有限公司
印　刷	天津久佳雅创印刷有限公司
规　格	210mm×285mm　16 开本　8.125 印张　250 千字
版　次	2023 年 10 月第 1 版　2023 年 10 月第 1 次印刷
定　价	89.00 元

《中国水旱灾害防御公报》编委会

Editorial Board of *China Flood and Drought Disaster Prevention Bulletin*

《中国水旱灾害防御公报》编辑部

Editorial Office of *China Flood and Drought Disaster Prevention Bulletin*

CONTENTS 目录

推动新阶段水利高质量发展
大藤峡水利枢纽

2022 年，我国极端天气事件频发，洪水、干旱、咸潮交叠并发，历史罕见，珠江、辽河等流域发生大洪水，长江流域、珠江流域东江和韩江等发生严重气象水文干旱。水利部深入学习贯彻党的二十大精神，全面落实习近平总书记关于防汛抗旱重要指示批示精神和党中央、国务院决策部署，始终把保障人民群众生命财产安全放在第一位，锚定“人员不伤亡、水库不垮坝、重要堤防不决口、重要基础设施不受冲击”和确保城乡供水安全的目标，组织指导各级水利部门把防汛抗旱作为重大政治责任和头等大事，树牢底线思维、极限思维，坚持防汛抗旱“两手抓”，科学精准调度水工程，抓实抓细各项防范应对措施，全力以赴打赢了水旱灾害防御硬仗。

注

（1）本公报未包括香港特别行政区、澳门特别行政区和台湾省统计数据，新疆生产建设兵团统计数据计入新疆维吾尔自治区统计数据；

（2）本公报所采用的计量单位部分沿用水利统计惯用单位，未进行换算；

（3）本公报数据来源于水利部、应急管理部，降水量数据依据水利部信息中心业务系统报汛数据统计，未注明来源的数据均来源于水利部，指标解释分别参阅《水旱灾害防御统计调查制度（试行）（2021）》《自然灾害情况统计调查制度（2020）》。

In 2022, extreme weather events of floods, droughts and saltwater intrusion occurred frequently and concurrently in China, rare in history, with major floods in the Pearl River and Liaohe River basins, and severe meteorological and hydrological droughts in the Yangtze River basin, and Dongjiang and Hanjiang Rivers that drain to the Pearl River basin. The Ministry of Water Resources (hereinafter MWR) thoroughly and forcefully executed the spirit of the 20th National Congress of the Communist Party of China, the key instructions of President Xi Jinping on flood prevention and drought relief, and the decisions and deployment of the CPC Central Committee and the State Council, put the safety of people's lives and property in the first place, kept up the goal of "no casualties, no dams collapses, no breach of important embankments, and no shocks on important infrastructure" and secured urban and rural water supply. Water resources departments at all levels were commanded and guided by the Ministry to tighten up the nerves against flood and drought disasters, think to the details and prepare for the worst, lay equal emphases on flood prevention and drought relief, scientifically and accurately dispatch water projects, implement various preemptive and preventive measures in a down-to-earth manner. The fight against flood and drought disasters has been tough and ended in success thanks to the all-out efforts.

Note

(1) The data in this Bulletin does not include statistics of the Hong Kong Special Administrative Region (SAR), the Macao Special Administrative Region (SAR) and Taiwan, and the statistics of the Xinjiang Production and Construction Corps is included in the statistics of the Xinjiang Uygur Autonomous Region;

(2) The units of measurement used in this Bulletin conform to what are customarily used in water conservancy and are not converted;

(3) The data in this Bulletin are from the Ministry of Water Resources (MWR) and the Ministry of Emergency Management (MEM), the data on precipitation are drawn upon the flood data reported on the service platform of MWR Water Resources Information Center, the data that do not indicate the source are all from the MWR, and interpretations on the indicators can be found in *Statistical Investigation System for Flood and Drought Disaster Prevention (Trial)(2021)* and *Statistical Investigation System for Natural Disasters (2020)*, respectively.

1 雨水情

RAINFALL AND WATER REGIME

1.1 雨情

2022 年，全国共发生 46 次强降雨过程，全国平均降水量 596 毫米，较常年（625 毫米）偏少 5%，为 2012 年以来最少；长江流域大部、黄河流域西部南部、淮河流域西部南部、珠江流域西部、太湖流域、浙闽流域大部等较常年偏少 1 ～ 2 成，黄河流域中北部、淮河流域东北部、海河流域东南部、松辽流域南部等较常年偏多 3 ～ 7 成。2022 年 5—9 月，全国平均降水量 404 毫米，较常年同期（453 毫米）偏少 11%，为 1973 年以来最少；长江流域中部东部、淮河流域西南部、太湖流域等较常年同期偏少 3 ～ 5 成，黄河流域北部、淮河流域东北部、海河流域东南部、松辽流域南部等较常年同期偏多 3 ～ 7 成。

2022 年，全国雨情总体有 3 个特点。

空间上南北多中间少。东北中部南部、华北南部、黄淮北部、华南南部等地部分地区降水量较常年偏多 2 成至 1 倍；江南、江淮、黄淮西部、西南等地部分地区降水量较常年偏少 2 ～ 5 成。辽宁省、吉林省、广东省年降水量较常年分别偏多 35%、27%、13%，分别为 1961 年有完整序列资料以来第 4 多、第 1 多、第 9 多；湖北省、贵州省、河南省年降水量较常年分别偏少 21%、20%、19%。

时间上前多后少。1—6 月降水量较常年同期偏多，尤其是 1—3 月降水量较常年同期明显偏多，1 月、2 月、3 月分别偏多 34%、33%、20%。7—10 月和 12 月降水量较常年同期偏少，尤其是 7—9 月降水量较常年同期明显偏少，7 月、8 月、9 月分别偏少 20%、23%、22%。

1.1 Rainfall

In 2022, 46 heavy rainfall processes occurred in China. The national average annual precipitation was 596 mm, 5% less than normal (625 mm) and the lowest since 2012. The most of the Yangtze River basin, the west and south of the Yellow River basin, the west and south of the Huaihe River basin, the west of the Pearl River basin, the Taihu Lake basin, and the most of the river basins in Zhejiang-Fujian provinces received 10%-20% less rainfall than normal. The central and northern parts of the Yellow River basin, the northeast of the Huaihe River basin, the southeast of the Haihe River basin, and the south of the Songhua-Liaohe River basin received 30%-70% more rainfall than normal. From May to September 2022, the national average precipitation was 404 mm, 11% less than normal over the same period (453 mm) and the lowest since 1973; the center and east of the Yangtze River basin, the southwest of the Huaihe River basin, and the Taihu Lake basin received 30%-50% less rainfall than normal over the same period, and the north of the Yellow River basin, the northeast of the Huaihe River basin, the southeast of the Haihe River basin, and the south of the Songhua-Liaohe River basin received 30%-70% more rainfall than normal over the same period.

In general, rainfall in 2022 took on the following three characteristics:

Spatially: more precipitation in the north and the south, and less in the central area.The precipitation in the central and southern parts of Northeast China, the south of North China, the north of Huanghuai (between the Yellow River and the Huaihe River) Plain, the south of South China, etc. was 20%-100% more than normal. Some areas of Jiangnan (areas south of the middle-lower Yangtze), Jianghuai (between the Yangtze and the Huaihe River), the west of Huanghuai, and Southwest China received 20%-50% less precipitation than normal. The annual precipitation in Liaoning Province, Jilin Province and Guangdong Province was 35%, 27% and 13% more than normal, respectively, which was the fourth, first and ninth highest since the complete sequence data was available in 1961. The annual precipitation in Hubei Province, Guizhou Province and Henan Province was 21%, 20% and 19% less than normal, respectively.

Temporally: more precipitation in the first half of the year and less in the second half. It was wetter than normal during January to June. In particular, precipitation was 34%, 33% and 20% more than its normal in January, February and March of 2022, respectively. It was drier than normal from July to October and in December. In particular, precipitation was 20%, 23% and 22% less than its normal in July, August and September, respectively.

南方夏秋降雨严重偏少。7 月至 11 月上旬，南方地区持续高温少雨，长江流域累计面雨量 282 毫米，较常年同期（523 毫米）偏少 46%，为 1961 年有完整序列资料以来第 1 少，其中鄱阳湖、洞庭湖、长江下游等地偏少 5～7 成；浙闽地区累计面雨量 221 毫米，较常年同期（543 毫米）偏少 59%，为 1961 年有完整序列资料以来第 1 少；珠江流域西江水系累计面雨量 301 毫米，较常年同期（496 毫米）偏少 39%，为 1961 年有完整序列资料以来第 2 少；太湖流域累计面雨量 312 毫米，较常年同期（504 毫米）偏少 38%，为 1961 年有完整序列资料以来第 5 少。

1.2 水情

1.2.1 江河径流量

2022 年，全国地表水径流量 25971.8 亿立方米，较多年平均值（26556.0 亿立方米）偏少 2.2%。全国主要江河径流量较常年总体偏少，其中，长江大通站径流量 7712.0 亿立方米，较常年偏少 2 成；黄河花园口站径流量 323.5 亿立方米，较常年偏少 1 成；淮河鲁台子站径流量 105.9 亿立方米，较常年偏少 5 成；海河流域拒马河张坊站径流量 1.8 亿立方米，较常年偏少 5 成；珠江流域西江梧州站径流量 2066.4 亿立方米，与常年持平，北江飞来峡站径流量 489.7 亿立方米，较常年偏多 4 成，东江博罗站径流量 189.1 亿立方米，较常年偏少 2 成；松花江佳木斯站径流量 749.3 亿立方米，较常年偏多 5 成；辽河铁岭站径流量 91.4 亿立方米，较常年偏多 2 倍。

注 松花江、辽河、海河、黄河流域径流量统计时段划分：汛前（1—5 月）、汛期（6—9 月）、汛后（10—12 月）；淮河、长江、珠江流域径流量统计时段划分：汛前（1—4 月）、汛期（5—9 月）、汛后（10—12 月）；太湖暂不做径流量统计。地表水径流量多年平均值为 1956—2016 年地表水径流量平均值。水情数据为实测数据。

The summer and autumn rainfall in the south was severely lean. From July to early November, the southern region had been subject to hot and dry days, and the cumulative rainfall over the Yangtze River basin was 282 mm, 46% less than normal (523 mm) and the least since complete sequence data was available in 1961. In particular, Poyang Lake, Dongting Lake and the lower Yangtze received 5%-7% less than normal. The cumulative rainfall over the Zhejiang-Fujian region was 221 mm, 59% less than normal over the same period (543 mm) and the lowest since complete sequence data was available in 1961. The cumulative rainfall over the Xijiang River system in the Pearl River basin was 301 mm, 39% less than normal over the same period (496 mm) and the second lowest since complete sequence data was available in 1961. The cumulative rainfall over the Taihu Lake basin was 312 mm, 38% less than normal over the same period (504 mm) and the fifth lowest since complete sequence data was available in 1961.

1.2 Water Regime

1.2.1 River discharge

In 2022, the national surface water runoff was 2,597.18 billion m^3, 2.2% less than the multi-year average (2,655.60 billion m^3). The discharge of major rivers was less than normal, among which the discharge at the Datong Station of the Yangtze River was 771.20 billion m^3, 20% less than normal. The discharge at the Huayuankou Station of the Yellow River was 32.35 billion m^3, 10% less. The discharge at the Lutaizi Station of the Huaihe River was 10.59 billion m^3, 50% less. The discharge at the Zhangfang Station of the Juma River (draining to the Haihe River basin) was 180 million m^3, 50% less. In the Pearl River basin, the discharge at the Wuzhou Station on Xijiang River was 206.64 billion m^3, the same as normal; the discharge at the Feilaixia Station on Beijiang River was 48.97 billion m^3, 40% more than normal; the discharge at Boluo Station on Dongjiang River was 18.91 billion m^3, 20% less. The discharge at the Jiamusi Station on the Songhua River basin was 74.93 billion m^3, 50% more than normal. The discharge at the Tieling Station on Liaohe River basin was 9.14 billion m^3, twice as much as normal.

Note For statistics of river discharges in the Songhua, the Liaohe, the Haihe, and the Yellow River basins: pre-flood period (January-May), flood period (June-September), post-flood period (October-December); for statistics of river discharges in the Huaihe, the Yangtze and the Pearl River basin: pre-flood period (January-April), flood period (May-September), post-flood period (October-December); No statistics of discharges in the Taihu Lake are prepared. The multi-year average of surface water runoff is the average surface water runoff from 1956 to 2016. The water data is measured data.

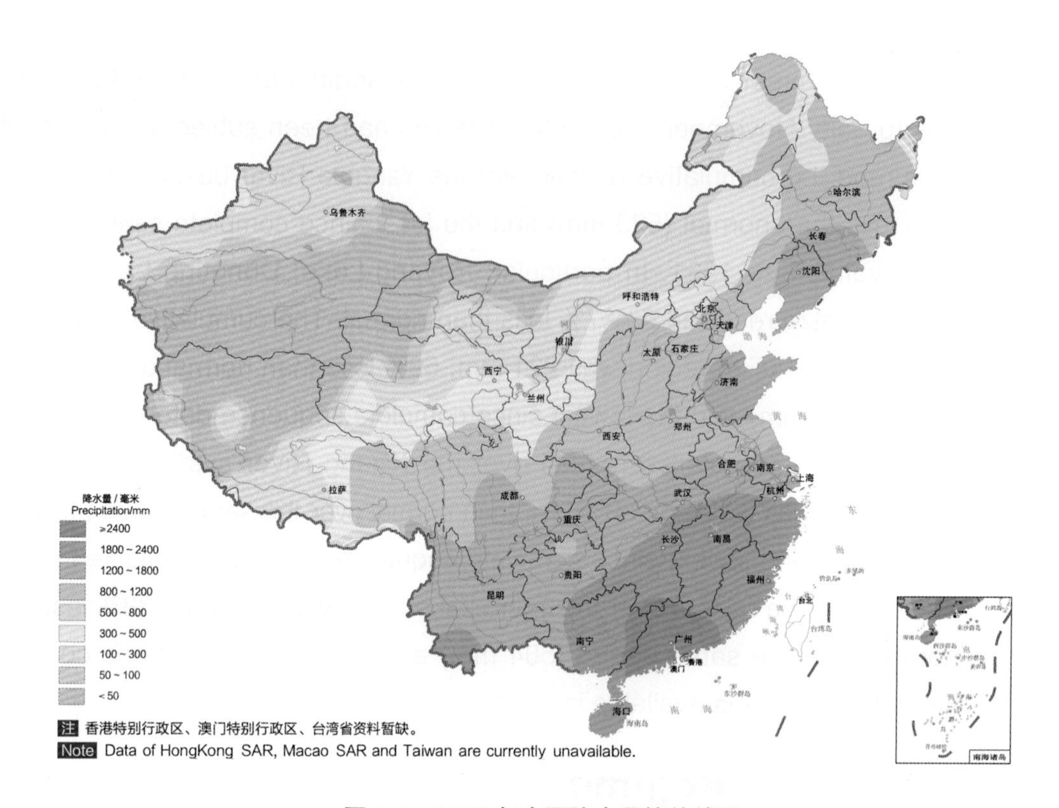

图 1-1　2022 年全国降水量等值线图
Figure 1-1　Isogram of national precipitation in 2022

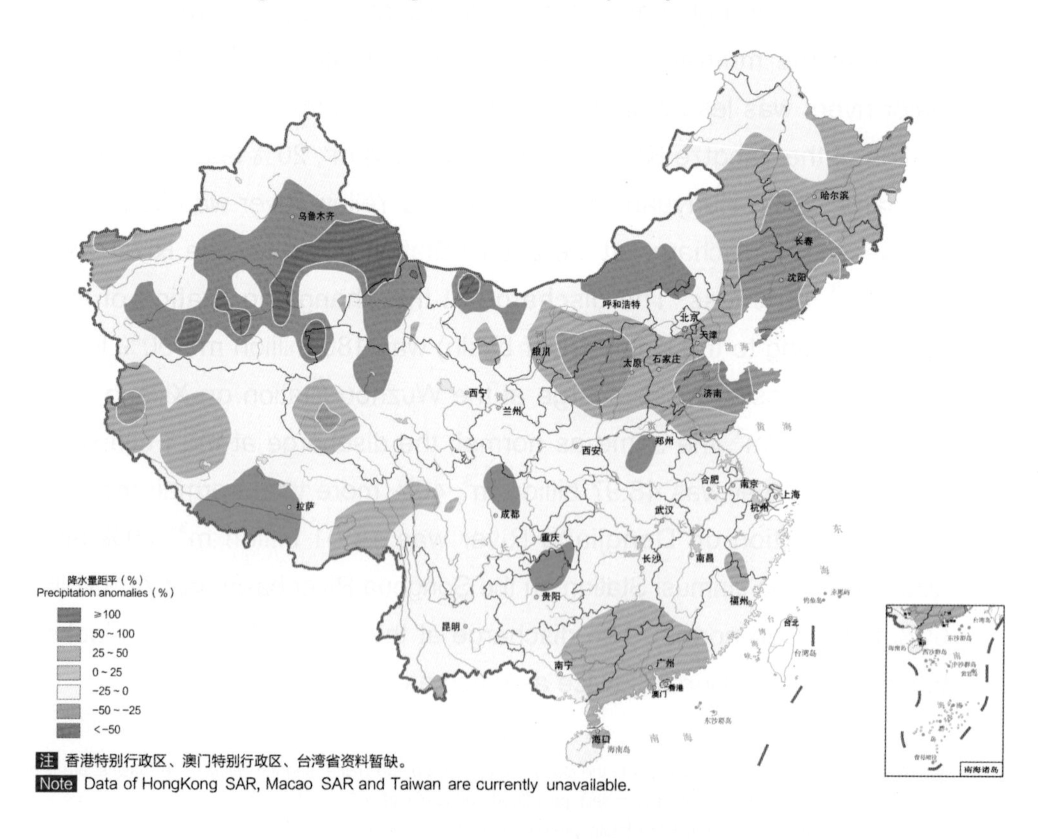

图 1-2　2022 年全国降水量距平图
Figure 1-2　National precipitation anomalies in 2022

乌鲁木齐	Urumqi
拉萨	Lhasa
西宁	Xining
银川	Yinchuan
兰州	Lanzhou
呼和浩特	Hohhot
哈尔滨	Harbin
长春	Changchun
沈阳	Shenyang
北京	Beijing
天津	Tianjin
石家庄	Shijiazhuang
太原	Taiyuan
济南	Jinan
郑州	Zhengzhou
西安	Xi'an
武汉	Wuhan
南京	Nanjing
上海	Shanghai
杭州	Hangzhou
合肥	Hefei
南昌	Nanchang
福州	Fuzhou
长沙	Changsha
重庆	Chongqing
成都	Chengdu
贵阳	Guiyang
广州	Guangzhou
昆明	Kunming
南宁	Nanning
海口	Haikou
澳门	Macao
香港	Hong Kong
台北	Taibei
海南岛	Hainan Dao
台湾岛	Taiwan Dao
钓鱼岛	Diaoyu Dao
赤尾屿	Chiwei Yu
渤海	Bo Hai
黄海	Yellow Sea
东海	East China Sea
南海	South China Sea
台湾海峡	Taiwan Strait
南海诸岛	Nanhai Islands
东沙群岛	Dongsha Qundao
南沙群岛	Nansha Qundao
中沙群岛	Zhongsha Qundao
西沙群岛	Xisha Qundao
黄岩岛	Huangyan Dao
曾母暗沙	Zengmu Ansha

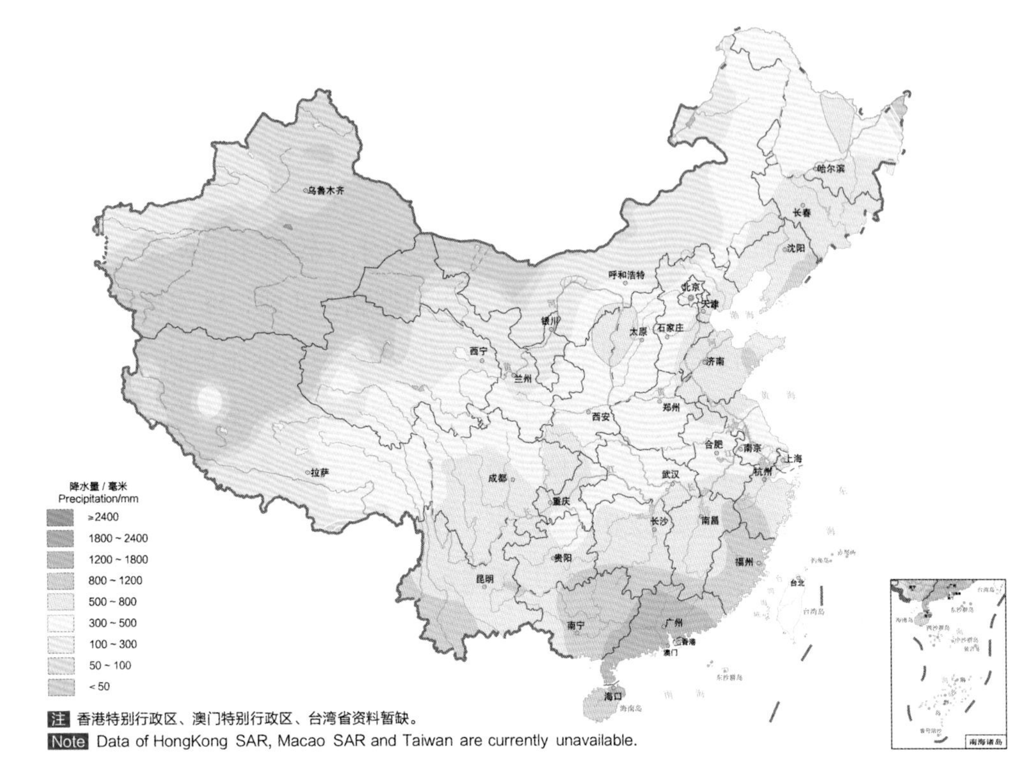

图 1-3 2022 年 5—9 月全国降水量等值线图

Figure 1-3 Isogram of precipitation from May to September 2022

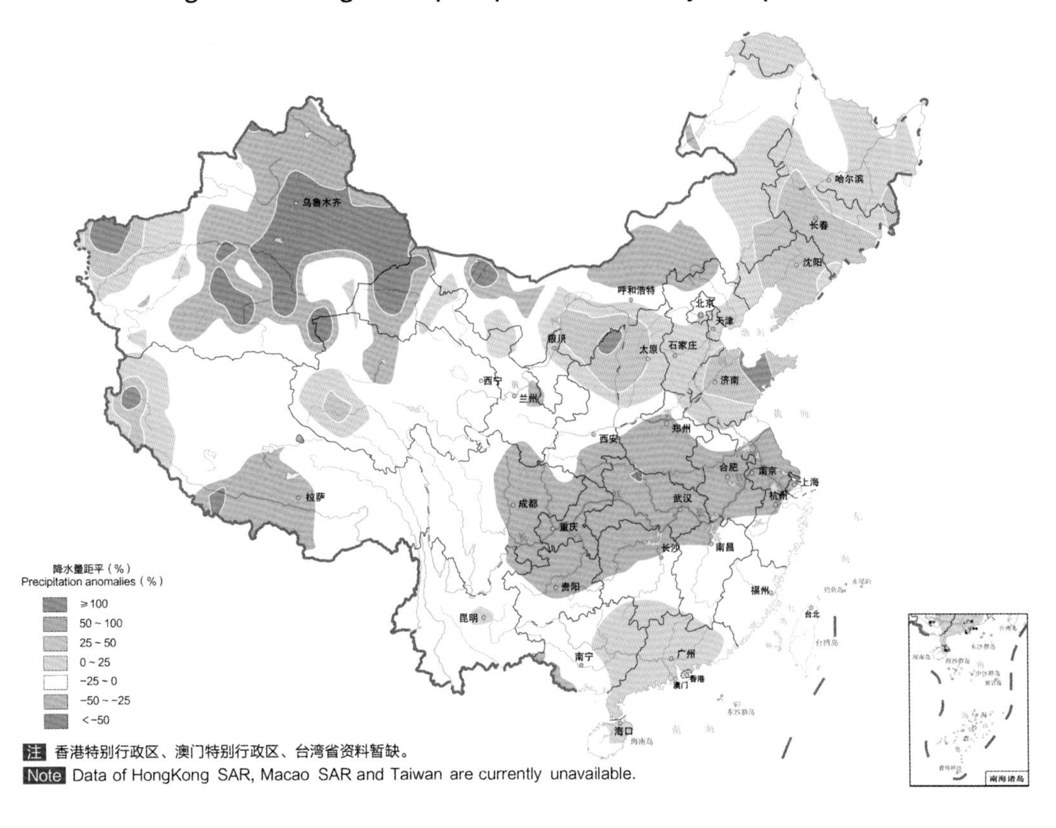

图 1-4 2022 年 5—9 月全国降水量距平图

Figure 1-4 Precipitation anomalies from May to September 2022

汛前，长江大通站径流量 2294.0 亿立方米，较常年同期偏多 2 成；黄河花园口站径流量 141.0 亿立方米，较常年同期偏多 4 成；淮河鲁台子站径流量 72.2 亿立方米，较常年同期偏多 7 成；海河流域拒马河张坊站径流量 0.7 亿立方米，较常年同期偏多 3 成；珠江流域西江梧州站径流量 852.5 亿立方米，较常年同期偏多 7 成，北江飞来峡站径流量 121.8 亿立方米，较常年同期偏少 2 成，东江博罗站径流量 52.0 亿立方米，较常年同期偏少 3 成；松花江佳木斯站径流量 178.8 亿立方米，较常年同期偏多 6 成；辽河铁岭站径流量 8.4 亿立方米，较常年同期偏多 8 成。

汛期，长江大通站径流量 4562.0 亿立方米，较常年同期偏少 3 成；黄河花园口站径流量 143.1 亿立方米，较常年同期偏少 1 成；淮河鲁台子站径流量 24.6 亿立方米，较常年同期偏少 8 成；海河流域拒马河张坊站径流量 0.8 亿立方米，较常年同期偏少 6 成；珠江流域西江梧州站径流量 1062.5 亿立方米，较常年同期偏少 2 成，北江飞来峡站径流量 344.8 亿立方米，较常年同期偏多 1.1 倍，东江博罗站径流量 108.0 亿立方米，与常年同期持平；松花江佳木斯站径流量 466.4 亿立方米，较常年同期偏多 5 成；辽河铁岭站径流量 73.0 亿立方米，较常年同期偏多 2.3 倍。

汛后，长江大通站径流量 856.0 亿立方米，较常年同期偏少 5 成；黄河花园口站径流量 39.4 亿立方米，较常年同期偏少 5 成；淮河鲁台子站径流量 9.1 亿立方米，较常年同期偏少 7 成；海河流域拒马河张坊站径流量 0.3 亿立方米，较常年同期偏少 5 成；珠江流域西江梧州站径流量 151.4 亿立方米，较常年同期偏少 5 成，北江飞来峡站径流量 23.1 亿立方米，较常年同期偏少 4 成，东江博罗站径流量 29.1 亿立方米，较常年同期偏少 2 成；松花江佳木斯站径流量 104.1 亿立方米，与常年同期持平；辽河铁岭站径流量 10.0 亿立方米，较常年同期偏多 2 倍。

Before the flood season, the discharge at the Datong Station on the Yangtze was 229.40 billion m^3, 20% more than normal over the same period. The Huayuankou Station on the Yellow River had a discharge of 14.10 billion m^3, 40% more than normal over the same period. The discharge at Lutaizi Station on the Huaihe was 7.22 billion m^3, 70% more than normal over the same period. The discharge at Zhangfang Station on the Juma (draining to the Haihe River basin) was 70 million m^3, 30% more than normal over the same period. In the Pearl River basin, the discharge at Wuzhou Station on the Xijiang was 85.25 billion m^3, 70% more than normal over the same period; the discharge at Feilaixia Station on the Beijiang was 12.18 billion m^3, 20% less than normal over the same period of the year; and the discharge at Boluo Station on the Dongjiang was 5.20 billion m^3, 30% less than normal over the same period. The discharge at Jiamusi Station on the Songhua was 17.88 billion m^3, 60% more than normal over the same period. The discharge at Tieling Station on the Liaohe was 840 million m^3, 80% more than normal over the same period.

During the flood season, the discharge at Datong Station on the Yangtze was 456.20 billion m^3, 30% less than normal over the same period. The Huayuankou Station on the Yellow River had a discharge of 14.31 billion m^3, 10% less than normal over the same period. The discharge at Lutaizi Station on the Huaihe was 2.46 billion m^3, 80% less than normal over the same period. The discharge at Zhangfang Station on the Juma (draining to the Haihe River basin) was 80 million m^3, 60% less than normal over the same period. In the Pearl River basin, the discharge at Wuzhou Station on the Xijiang was 106.25 billion m^3, 20% less than normal over the same period; the discharge at Feilaixia Station on the Beijiang was 34.48 billion m^3, 1.1 times more than normal over the same period; and the discharge at Boluo Station on the Dongjiang was 10.80 billion m^3, the same as normal. The discharge at Jiamusi Station on the Songhua was 46.64 billion m^3, 50% more than normal over the same period. The discharge at Tieling Station on the Liaohe was 7.30 billion m^3, 2.3 times more than normal over the same period.

After the flood season, the discharge at Datong Station on the Yangtze was 85.60 billion m^3, 50% less than normal over the same period. The discharge at the Huayuankou Station on the Yellow River was 3.94 billion m^3, 50% less than normal over the same period. The discharge at Lutaizi Station on the Huaihe was 910 million m^3, 70% less than normal over the same period. The discharge at Zhangfang Station on the Juma (draining to the Haihe River basin) was 30 million m^3, 50% less than normal over the same period. In the Pearl River basin, the discharge at Wuzhou Station on the Xijiang was 15.14 billion m^3, 50% less than normal over the same period; the discharge at Feilaixia Station on the Beijiang was 2.31 billion m^3, 40% less than normal over the same period; and the discharge at Boluo Station on the Dongjiang was 2.91 billion m^3, 20% less than normal over the same period. The discharge at Jiamusi Station on the Songhua was 10.41 billion m^3, the same as normal over the same period. The discharge at Tieling Station on the Liaohe was 1 billion m^3, twice as much as normal the same period.

1.2.2 水库蓄水

6 月 1 日，纳入水利部日常统计范围的 6920 座水库蓄水量（以下简称统计水库蓄水量）3957.2 亿立方米，较常年同期偏多 17%。其中，709 座大型水库蓄水量 3473.4 亿立方米，较常年同期偏多 17%；2980 座中型水库蓄水量 406.2 亿立方米，较常年同期偏多 7%。

10 月 1 日，统计水库蓄水量 4381.5 亿立方米，较 6 月 1 日增加 11%，较常年同期偏少 3%。其中，709 座大型水库蓄水量 3970.0 亿立方米，较 6 月 1 日增加 14%，较常年同期偏少 3%；2980 座中型水库蓄水量 380.5 亿立方米，较 6 月 1 日减少 6%，较常年同期偏少 5%。

年末（2023 年 1 月 1 日），统计水库蓄水量 4375.0 亿立方米，较 1 月 1 日减少 6%，较常年同期偏少 9%。其中，709 座大型水库蓄水量 3966.3 亿立方米，较 1 月 1 日减少 6%，较常年同期偏少 9%；2980 座中型水库蓄水量 378.7 亿立方米，较 1 月 1 日减少 6%，较常年同期偏多 1%。

表 1-1 2022 年统计水库蓄水量情况

Table 1-1 Water storage of the daily-reporting reservoirs in 2022

时间 Date	6920 座水库蓄水量 / 亿立方米 Storage in 6,920 reservoirs / 100 million m^3			
	709 座大型水库 709 large reservoirs	2980 座中型水库 2,980 medium-sized reservoirs	3231 座小型水库 3,231 small reservoirs	合计 Total
1 月 1 日 January 1	4202.7	403.4	38.0	4644.1
6 月 1 日 June 1	3473.4	406.2	77.6	3957.2
10 月 1 日 October 1	3970.0	380.5	31.0	4381.5
年末 Year-end	3966.3	378.7	30.0	4375.0

注 年末数据指 2022 年 1 月 1 日统计数据。

Note Year-end data as of January 1, 2022.

1.2.2 Reservoir storage

On June 1, the 6,920 reservoirs that report daily statistics to MWR (hereinafter referred to as the storage of daily-reporting reservoirs) had a total storage of 395.72 billion m^3, 17% more than normal over the same period. Among them, 709 large reservoirs had a water storage of 347.34 billion m^3, 17% more than normal over the same period; and the 2,980 medium-sized reservoirs had a total storage of 40.62 billion m^3, 7% more than normal over the same period.

On October 1, the storage of daily-reporting reservoirs was 438.15 billion m^3, 11% more than that on June 1 and 3% less than normal over the same period. Among them, the 709 large reservoirs stored 397.00 billion m^3, 14% more than that on June 1 and 3% less than normal over the same period; and the 2,980 medium-sized reservoirs stored 38.05 billion m^3, 6% less than that on June 1 and 5% less than normal over the same period.

At the end of the year (January 1, 2023), the storage of daily-reporting reservoirs was 437.50 billion m^3, 6% less than that on January 1, 2022 and 9% less than normal over the same period. Among them, the 709 large reservoirs stored 396.63 billion m^3, 6% less than that on January 1, 2022 and 9% less than normal over the same period. The 2,980 medium-sized reservoirs stored 37.87 billion m^3, 6% less than that on January 1, 2022 and 1% more than normal over the same period.

2 洪涝灾害防御

FLOOD DISASTER PREVENTION

2.1 汛情

2022年3月17日，我国进入汛期，较多年平均入汛日期偏早15天。2022年，全国主要江河共发生10次编号洪水，共有626条河流发生超警戒洪水，其中90条河流发生超保证洪水、27条河流发生超历史实测记录洪水。长江流域（片）有136条河流发生超警戒洪水，其中秋浦河等23条河流发生超保证洪水，乐安河等9条河流发生超历史实测记录洪水；黄河流域（片）有49条河流发生超警戒洪水，其中新疆塔里木河等13条河流发生超保证洪水，泾河、五当沟等5条河流发生超历史实测记录洪水；淮河流域（片）有10条河流发生超警戒洪水，其中小清河发生超保证洪水，小清河、清洋河2条河流发生超历史实测记录洪水，沭河发生1次编号洪水；海河流域（片）有1条河流发生超警戒洪水；珠江流域（片）发生2次流域性较大洪水，有292条河流发生超警戒洪水，其中桂江等14条河流发生超保证洪水，连江、灵渠等5条河流发生超历史实测记录洪水，西江、北江、韩江共发生8次编号洪水，北江发生1915年以来最大洪水；松辽流域（片）有55条河流发生超警戒洪水，其中伊通河等15条河流发生超保证洪水，鸭绿江、绕阳河等4条河流发生超历史实测记录洪水，辽河发生1次编号洪水；太湖流域（片）有83条河流发生超警戒洪水，其中建溪等24条河流发生超保证洪水，姚江、松溪2条河流发生超历史实测记录洪水。黄河、黑龙江等北方河流凌情平稳。

注 全国及各流域（片）汛情数据来源于水利部信息中心；河流超警戒（超保证、超历史）包括水位超警戒（超保证、超历史）或流量超警戒（超保证、超历史）。

2.1 Floods

The 2022 flood season in China began on March 17, 15 days earlier than the multi-year average start date. In 2022, a total of 10 numbered floods occurred in major rivers across the country. A total of 626 rivers swelled above the warning level, of which 90 rivers experienced floods that exceed the guaranteed levels and 27 rivers experienced record-breaking floods. In the Yangtze River basin/region, 136 rivers experienced floods beyond the warning levels; among them, 23 rivers, including the Qiupu River, had floods beyond the guaranteed levels, and 9 rivers, including the Le'an River, had record-breaking floods. In the Yellow River basin/region, 49 rivers experienced floods beyond the warning levels; among them, 13 rivers, including the Tarim River in Xinjiang, had floods beyond the guaranteed levels, and 5 rivers, including the Jinghe River and the Wudanggou had record-breaking floods. In the Huaihe River basin/region, 10 rivers experienced floods beyond the warning levels; among them, the Xiaoqing River had floods beyond the guaranteed level, the Xiaoqing River and the Qingyang River had record-breaking floods, and the Shuhe River had one numbered flood. In the Haihe River basin/region, one river experienced floods beyond the warning level. In the Pearl River basin/region, 2 large basin-wide floods occurred and 292 rivers had floods beyond the warning levels; among them, 14 rivers, including the Guijiang River, had floods beyond the guaranteed levels, 5 rivers such as the Lianjiang and the Lingqu had record-breaking floods, 8 numbered floods occurred in the Xijiang, Beijiang and Hanjiang rivers, and the Beijiang withstood the largest flood since 1915. In the Songhua-Liaohe River basin/region, 55 rivers experienced floods beyond the warning levels; among them, 15 rivers such as the Yitong River had floods beyond the guaranteed levels, 4 rivers such as the Yalu and the Raoyang River had record-breaking floods, and one numbered flood occurred in the Liaohe River. In the Taihu Lake basin/region, 83 rivers experienced floods beyond the warning levels; among them 24 rivers such as the Jianxi had floods beyond the guaranteed levels, and the Yaojiang and the Songxi rivers had record-breaking floods. China's northern rivers, such as the Yellow River and Heilong River, safely passed the ice flood period.

Note Data on floods nationwide and by different water basins/regions are from the Water Resources Information Center, MWR. River floods that break the warning/guaranteed/maximum measured levels include those that break the warning/guaranteed/maximum measured water level and those that break the warning/guaranteed/maximum measured discharges.

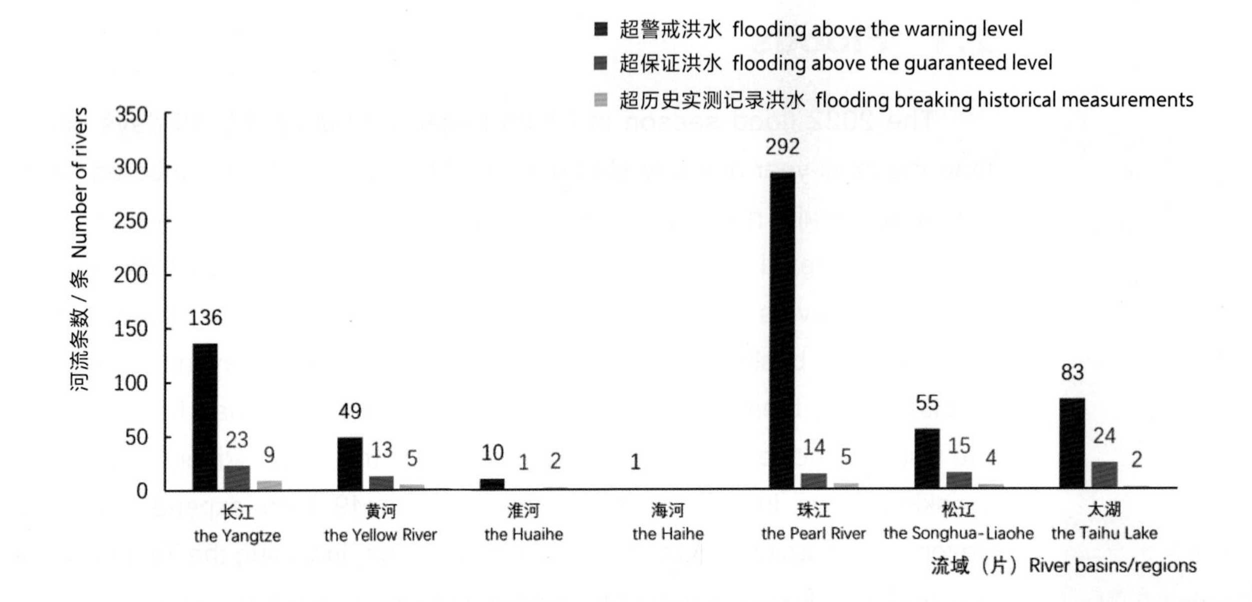

图 2-1　2022 年各流域（片）发生超警戒、超保证、超历史实测记录洪水河流条数

Figure 2-1　Number of rivers experiencing floods above the warning water level, above the guaranteed water level, and breaking historical measurements in the major river basins in 2022

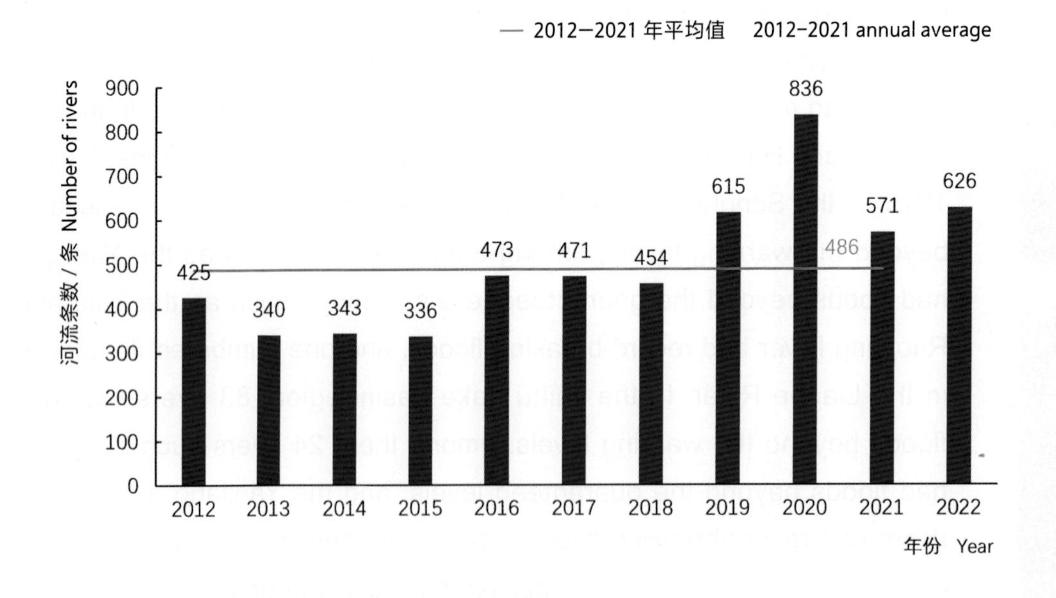

图 2-2　2012—2022 年全国发生超警戒洪水河流条数

Figure 2-2　Number of rivers experiencing floods above the warning level in China 2012−2022

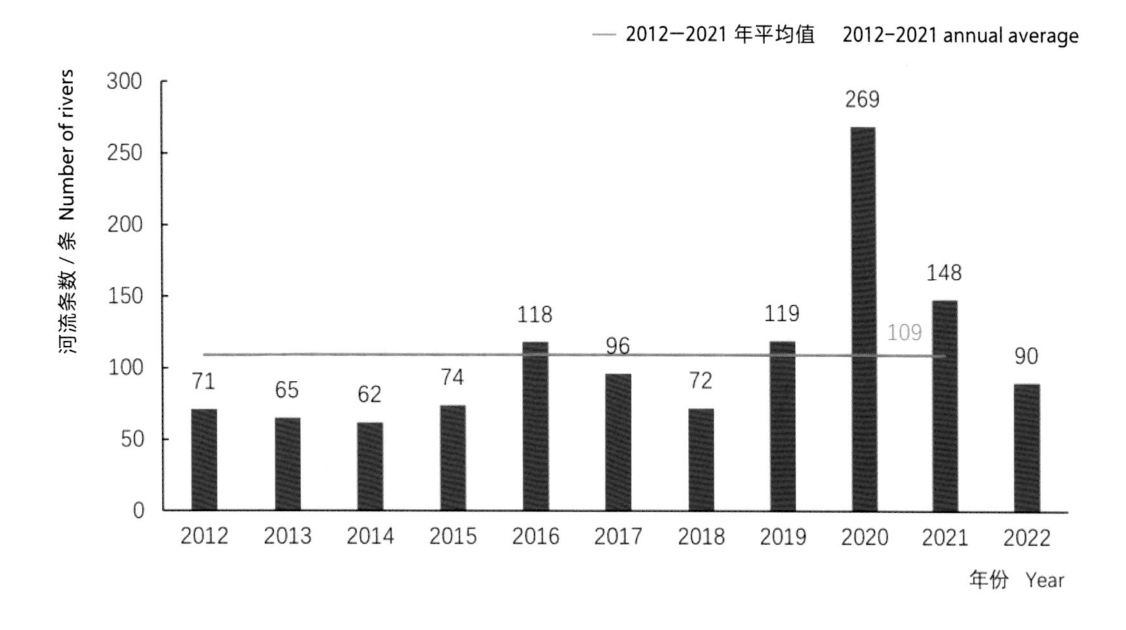

图 2-3 2012—2022 年全国发生超保证洪水河流条数
Figure 2-3 Number of rivers experiencing floods above the guaranteed level in China 2012-2022

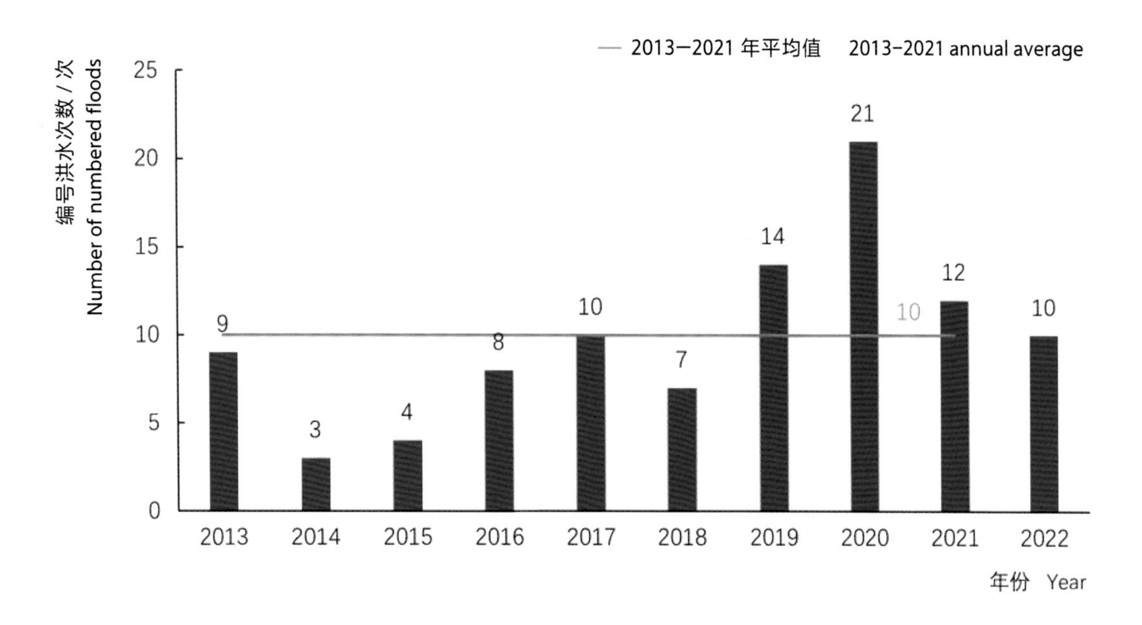

图 2-4 2013—2022 年全国主要江河发生编号洪水次数
Figure 2-4 Statistics of numbered floods in major rivers in China 2013-2022

2.2 汛情特点

2022 年，全国汛情总体有 3 个特点。

珠江流域编号洪水多、量级大。5 月下旬至 7 月上旬，珠江流域发生 2 次流域性较大洪水，西江、北江、韩江共发生 8 次编号洪水，北江发生 1915 年以来最大洪水，北江干流英德站水位最大，超警戒 9.97 米，北江支流连江、广西灵渠发生有实测记录以来最大洪水。

北方局地汛情罕见。松辽流域辽河发生 1995 年以来最大洪水，干流超警戒历时 47 天，支流绕阳河发生 1951 年以来最大洪水；新疆维吾尔自治区塔里木河干流持续超警戒 80 天，支流托什干河、阿克苏河融雪洪水发生时间分别较 2021 年提前 42 天、27 天；青海省那棱格勒河发生大洪水；内蒙古自治区黑河干流发生超保证洪水；陕西省泾河出现 1965 年有实测记录以来最高水位。

台风暴雨洪水影响范围广。2203 号台风“暹芭”在粤西登陆后穿过广东省、广西壮族自治区、湖南省，其残余环流继续北上影响黄淮、东北等地，共影响广东、广西、湖南、湖北、江西等 16 个省（自治区、直辖市），淮河干流、沂河、南四湖等河湖出现明显涨水过程。2212 号台风“梅花”为 1949 年以来第 3 个在我国 4 次登陆的台风，在浙江省、上海市登陆后又北上登陆山东省、辽宁省，共影响上海、浙江、江苏、山东、辽宁等 10 省（直辖市），浙江省姚江发生有实测记录以来最大洪水，山东省清洋河发生 1960 年有实测记录以来最大洪水。

2.2 Flood Characteristics

Floods that occurred in China in 2022 generally took on the following three characteristics:

The Pearl River basin was hit by many numbered floods with high floodwater.From late May to early July, two large basin-wide floods hit the Pearl River basin, eight numbered floods occurred in the Xijiang, Beijiang and Hanjiang rivers, the Beijiang experienced its largest flood since 1915, floodwater at the Yingde Station on the mainstream Beijiang exceeded the warning level by 9.97 m, and the Lianjiang (a tributary of the Beijiang) and the Lingqu Canal (in Guangxi Zhuang Autonomous Region) withstood floods that broke historical measurements.

Parts of North China saw floods that were rare in history.The Liaohe River in the Songhua-Liaohe River basin had its largest flood since 1995; the mainstream withstood floodwater beyond the warning levels for 47 days and its tributary — the Raoyang River experienced the largest flood since 1951. The mainstream Tarim River in Xinjiang Uygur Autonomous Region withstood floods above the warning levels for as long as 80 days, and the snowmelt floods in the tributaries — Toshken River and Aksu River — occurred 42 days and 27 days earlier than in 2021, respectively. The Nalenggele River in Qinghai Province experienced large floods. The mainstream Heihe River in Inner Mongolia Autonomous Region experienced floods above the guaranteed levels. The water level in the Jinghe River in Shaanxi Province registered a new high since measurements began in 1965.

Rainstorms and floods triggered by typhoons affected a wide area.Typhoon "Chaba" (No. 2203) swept through Guangdong, Guangxi and Hunan after landing in western Guangdong, and its residual circulation continued to move north to affect the Huanghuai region and Northeast China, affecting 16 provinces/autonomous regions/municipalities including Guangdong, Guangxi, Hunan, Hubei and Jiangxi. As a result, the mainstream Huaihe River, the Yihe River, and the Nansi Lake swelled notably. Typhoon "Muifa" (No. 2212) was the third typhoon to make four landfalls in China in a year since 1949, hitting Zhejiang Province and Shanghai first before swirling north to Shandong and Liaoning provinces. It affected 10 provinces/municipalities including Shanghai, Zhejiang, Jiangsu, Shandong and Liaoning, resulting in record-breaking floods happening in the Yaojiang River in Zhejiang Province and also in the Qingyang River in Shandong Province (where River measurements began in 1960).

2.3 主要洪水过程

2022 年，全国主要江河先后发生 3 次主要洪水过程。

2.3.1 5—7 月珠江流域洪水

5 月下旬至 7 月上旬，珠江流域共发生 11 次强降雨过程，整体降雨历时近 50 天，累计面雨量 622 毫米，较常年同期偏多 4 成，其中北江、韩江流域累计面雨量均为 1961 年以来同期最多。强降雨主要发生在柳江、桂江、贺江、北江等流域中北部地区，暴雨落区高度重叠。西江先后发生 4 次编号洪水，北江先后发生 3 次编号洪水，韩江发生 1 次编号洪水。其中，北江第 2 号洪水期间，6 月 22 日 23 时中游干流代表站飞来峡水库出现最大入库流量 19900 立方米每秒，仅次于 1915 年（调查洪水洪峰流量 21500 立方米每秒），为北江 1915 年以来最大洪水。6 月 14 日 11 时 30 分北江干流石角站流量 12000 立方米每秒，6 月 14 日 12 时西江干流梧州站水位 21.61 米，超过警戒水位（18.50 米）3.11 米，相应流量 37300 立方米每秒，珠江流域发生流域性较大洪水。6 月 19 日 8 时西江干流梧州站水位 20.95 米，超过警戒水位（18.50 米）2.45 米，相应流量 34500 立方米每秒，6 月 19 日 12 时北江干流石角站流量 12000 立方米每秒，珠江流域再次发生流域性较大洪水。

西江第 3 号洪水经过广西梧州河段（6 月 14 日）
No.3 Flood of Xijing River passed through the Wuzhou Section in Guangxi (June 14)

2.3 Major Flood Processes

Below is the account of the three major flood processes in major rivers in 2022.

2.3.1 Flooding in the Pearl River basin from May to July

From late May to early July, a total of 11 heavy rainfall processes occurred in the Pearl River basin, the overall rainfall lasted for nearly 50 days, and the cumulative rainfall over the basin was 622 mm, 40% more than normal over the same period. In particular, the cumulative rainfall over the Beijiang and the Hanjiang river basins were the highest over the same period since 1961. Heavy rainfall mainly occurred in the central and northern parts of the Liujiang, Guijiang, Hejiang, and Beijiang river basins, and torrential rains repetitively lashing the same areas. There were four numbered floods in the Xijiang River, three numbered floods in the Beijiang River, and one numbered flood in the Hanjiang River. In particular, at 23:00 on June 22 during the No. 2 Flood in the Beijiang, Feilaixia station, a typical one on the mainstream, record a peak inflow of 19,900 m^3/s, which was second only to that in 1915 (the forensic flood peak flow was 21,500 m^3/s). It was the largest flood in Beijiang since 1915. At 11:30 on June 14, the Shijiao Station on the mainstream Beijiang measured a flow of 12,000 m^3/s; at 12:00 on June 14, the water level at Wuzhou Station on the mainstream Xijiang rose to 21.61 m, exceeding the warning water level (18.50 m) by 3.11 m, and the corresponding flow was 37,300 m^3/s. The Pearl River basin consequently experienced medium flood basin-wide. At 8:00 on June 19, the water level at Wuzhou Station on the mainstream Xijiang rose to 20.95 m, exceeding the warning water level (18.50 m) by 2.45 m, and the corresponding flow was 34,500 m^3/s; at 12:00 on June 19, the flow at Shijiao Station on the mainstream Beijiang reached 12,000 m^3/s. Another medium flooding hit the Pearl River basin consequently.

表 2-1　2022 年珠江流域编号洪水情况
Table 2-1　Numbered floods in the Pearl River basin in 2022

序号 No.	时间 Start time of numbering	洪水编号 Numbered floods	编号依据 Formation of numbering
1	5 月 30 日 11 时 11:00 May 30	西江 2022 年第 1 号洪水 No. 1 flood in the Xijiang River, 2022	西江上游龙滩水库入库流量 10900 立方米每秒 The inflow of Longtan Reservoir in the upper Xijiang in the Pearl River basin reached 10,900 m³/s
2	6 月 6 日 17 时 17:00 June 6	西江 2022 年第 2 号洪水 No. 2 flood in the Xijiang River, 2022	西江中游广西武宣站流量 25200 立方米每秒 The flow at the Wuxuan Station in the middle Xijiang in the Pearl River basin reached 25,200 m³/s
3	6 月 12 日 20 时 20:00 June 12	西江 2022 年第 3 号洪水 No. 3 flood in the Xijiang River, 2022	西江干流广西梧州站水位涨至 18.52 米，超过警戒水位 0.02 米 The water level at Wuzhou Station in Guangxi on the main stream Xijiang River rose to 18.52 m, exceeding the warning water level by 0.02 m
4	6 月 13 日 14 时 14:00 June 13	韩江 2022 年第 1 号洪水 No. 1 flood in the Hanjiang River, 2022	韩江广东三河坝站水位涨至 42.73 米，超过警戒水位 0.73 米，流量达到 4890 立方米每秒 The water level at Sanheba Station on the Hanjiang in Guangdong rose to 42.73 m, exceeding the warning water level by 0.73 m; and the flow reached 4,890 m³/s
5	6 月 14 日 11 时 30 分 11:30 June 14	北江 2022 年第 1 号洪水 No. 1 flood in the Beijiang River, 2022	北江干流广东石角站流量 12000 立方米每秒 The flow at Shijiao Station on the main stream Beijiang in Guangdong in the Pearl River basin reached 12,000 m³/s
6	6 月 19 日 8 时 08:00 June 19	西江 2022 年第 4 号洪水 No. 4 flood in the Xijiang River, 2022	西江干流广西梧州站水位复涨至 20.95 米，超警戒水位 2.45 米 The water level at Wuzhou Station on the main stream Xijiang in Guangxi rose again to 20.95 m, 2.45 m above the warning water level
7	6 月 19 日 12 时 12:00 June 19	北江 2022 年第 2 号洪水 No. 2 flood in the Beijiang River, 2022	北江干流广东石角站流量 12000 立方米每秒 The flow at Shijiao Station on the main stream Beijiang in Guangdong reached 12,000 m³/s
8	7月5日7时35分 7:35 July 5	北江 2022 年第 3 号洪水 No. 3 flood in the Beijiang River, 2022	北江干流广东石角站流量 12000 立方米每秒 The flow at Shijiao Station on the main stream Beijiang in Guangdong reached 12,000 m³/s

2.3.2 6—8 月辽河流域洪水

6—8 月，辽河流域共发生 12 次主要降雨过程，降雨量 427 毫米，较常年同期偏多 3 成，主雨区均位于东辽河和辽河干流区域。受持续降雨影响，辽河流域 17 条河流发生超警戒洪水，其中 4 条河流发生超保证洪水，辽河干流福德店至盘山闸河段水位全线超警戒 47 天，最大超警戒幅度 0.70 ～ 1.82 米；东辽河干流水位全线超警戒 33 天，最大超警戒幅度 0.31 ～ 1.10 米。7 月 17 日 11 时辽河铁岭站水位达到警戒水位（60.22 米），形成辽河 2022 年第 1 号洪水；7 月 18 日铁岭站洪峰水位 60.53 米，为辽河 1995 年以来最大洪水；7 月 28 日支流绕阳河上游发生 1951 年有实测记录以来最大洪水，韩家杖子站超保证水位（112.03 米）1.31 米，相应流量 2590 立方米每秒，洪水重现期超 50 年。

辽河干流福德店河段洪水（7 月 18 日）
Flooding in the Fudedian section on the mainstream Liaohe River (July 18)

2.3.3 6 月下旬沂沭泗流域洪水

6 月 26—28 日，受低涡切变线和低空急流共同影响，沂沭泗流域发生强降雨过程，累计面雨量 118.7 毫米，其中临沂以上 135 毫米、大官庄以上 185 毫米、南四湖区 125 毫米、邳苍区 112 毫米、新沂河区 93 毫米。受强降雨影响，沂河、沭河、新沂河、南四湖、骆马湖等主要河湖出现明显涨水过程，6 月 27 日 12 时 30 分，沭河重沟站流量 2170 立方米每秒，形成沭河 2022 年第 1 号洪水。

2022 年，受局地暴雨影响，江西省乐安河，福建省建溪支流松溪，陕西省泾河，山东省北大沙河、小清河，吉林省鸭绿江上游等河流发生超历史实测记录洪水。

2.3.2 Flooding in the Liaohe River basin from June to August

From June to August, a total of 12 major rainfall processes occurred in the Liaohe River basin. The rainfall amounted to 427 mm, 30% more than normal over the same period; and the main rainfall areas were all located in the Dongliao River and the mainstream Liaohe River. Affected by the continuous rainfall, 17 rivers in the Liaohe River basin experienced floods above the warning levels, of which 4 rivers had floods above the guaranteed levels. The water level of the section from Fudedian to Panshanzha on the mainstream Liaohe stayed above the warning levels for 47 days, with the maximum exceedance ranging between 0.70-1.82 m. The mainstream Dongliao River experienced floods above the warning levels for 33 days, with the maximum exceedance ranging between 0.31-1.10 m. At 11:00 on July 17, the water level at Tieling Station on the Liaohe River reached the warning water level of 60.22 m, forming the No. 1 flood of the Liaohe River in 2022. On July 18, the peak water level at Tieling Station was 60.53 m, making the largest flood hitting the Liaohe River since 1995. On July 28, the largest flood occurred in the upper reaches of Raoyang River, a tributary of Liaohe River, since measurements began in 1951; and the water level at Hanjiazhangzi Station exceeded the guaranteed water level (112.03 m) by 1.31 m with a corresponding flow at 2,590 m^3/s. The flood's return period went beyond 50 years.

2.3.3 Flooding in Yihe-Shuhe-Sihe River basin in late June

From June 26 to 28, under the combined influence of the low vortex shear line and the low-altitude jet stream, a heavy rainfall process occurred in the Yihe-Shuhe-Sihe River basin and the cumulative rainfall was over the basin was 118.7 mm. In particular, the rainfall reached 135 mm in the area upstream from Linyi, 185 mm in the area upstream from Daguanzhuang, 125 mm in Nansi Lake area, 112 mm in Picang area and 93 mm in Xinyihe area. Affected by heavy rainfall, major rivers and lakes such as Yihe, Shuhe, Xinyihe, Nansi Lake, and Luoma Lake swelled notably, and at 12:30 on June 27, the flow at Chonggou Station on Shuhe River rated at 2,170 m^3/s, forming the No. 1 flood of Shuhe River in 2022.

In 2022, affected by local heavy rainfall, rivers such as the Le'an in Jiangxi Province, Songxi (a tributary of Jianxi River) in Fujian Province, the Jinghe in Shaanxi Province, the Beidasha and the Xiaoqing in Shandong Province, and the upper Yalu River in Jilin Province experienced floods that beat historical measurements.

2.4 洪涝灾情

2.4.1 基本情况

2022 年，全国 29 省（自治区、直辖市）发生不同程度洪涝灾害。因洪涝共有 3385.26 万人次受灾，比前 10 年的平均值下降 57.2%；171 人死亡失踪，比前 10 年的平均值下降 71.3%；3.13 万间房屋倒塌，比前 10 年的平均值下降 87.6%；农作物受灾面积 3413.73 千公顷，比前 10 年的平均值下降 54.3%，其中绝收面积 492.65 千公顷；直接经济损失 1288.99 亿元，占当年 GDP 的 0.11%，比前 10 年直接经济损失占当年 GDP 百分比的平均值下降 64.5%；各项指标均为近 10 年来最低。江西、福建、广东、广西、辽宁、湖南等 6 省（自治区）灾情较重，因洪涝直接经济损失 882.39 亿元，占全国的 68.5%。

2.4 Disasters and Losses

2.4.1 Summary

In 2022, flood disasters of varying degrees were borne by 29 provinces/autonomous regions/municipalities across China. A total of 33,852,600 person-times were affected by floods, down by 57.2% from the preceding decadal average; 171 people died or went missing, down by 71.3% from the preceding decadal average; 31,300 dwellings collapsed, down by 87.6% from the preceding decadal average; the affected cropland area was 3,413,730 ha, down by 54.3% from the preceding decadal average; the failed cropland area was 492,650 ha; the direct economic loss was 128.899 billion RMB, accounting for 0.11% of the annual GDP and down by 64.5% from the average of the direct economic loss as a percentage of GDP in the previous 10 years. All these indicators were the lowest in the past 10 years. The six provinces/autonomous regions of Jiangxi, Fujian, Guangdong, Guangxi, Liaoning and Hunan were the hardest hit: their direct economic losses attributable to floods billed 88.239 billion RMB and accounted for 68.5% of the national total.

表 2-2　2022 年全国因洪涝受灾人口、死亡人口、失踪人口及直接经济损失情况

Table 2-2　Population affected, deaths, missing persons and direct economic losses by floods in 2022

地区 Province	受灾人口 / 万人次 Affected population/ 10,000 person-times	死亡人口 / 人 Deaths/person	失踪人口 / 人 Missing persons/ person	直接经济损失 / 亿元 Direct economic losses /100 million RMB
全国 Nationwide	3385.26	143	28	1288.99
北京 Beijing	0.07			
天津 Tianjin				0.01
河北 Hebei	21.46			2.31
山西 Shanxi	113.77	7		25.27
内蒙古 Inner Mongolia	101.22	16	1	67.50
辽宁 Liaoning	280.09			126.40
吉林 Jilin	71.32			14.29
黑龙江 Heilongjiang	10.78	6	2	4.38
上海 Shanghai				
江苏 Jiangsu				
浙江 Zhejiang	35.31	1	1	28.41
安徽 Anhui	18.19			1.09
福建 Fujian	107.99			168.71
江西 Jiangxi	423.71	3		185.56
山东 Shandong	18.55	2		10.34
河南 Henan	46.78	1		4.21
湖北 Hubei	138.34			13.86
湖南 Hunan	422.31			109.92
广东 Guangdong	248.01	5		161.13
广西 Guangxi	483.44	6	2	130.67
海南 Hainan	0.15			0.02
重庆 Chongqing	52.76	1		5.61
四川 Sichuan	222.46	33	13	48.67
贵州 Guizhou	154.27	8	2	43.15
云南 Yunnan	175.83	8	1	29.74
西藏 Xizang	3.99			0.31
陕西 Shaanxi	117.90			36.58
甘肃 Gansu	85.73	13		53.73
青海 Qinghai	19.35	29	5	13.23
宁夏 Ningxia	8.17			1.72
新疆 Xinjiang	3.31	4	1	2.17

注 数据来源于应急管理部，空白表示无灾情，部分地区直接经济损失未完全包括水利工程设施直接经济损失。

Note The data come from the Ministry of Emergency Management, spaces in blank denote no such losses by floods, and the direct economic losses in some areas do not fully cover the direct economic losses by water engineering projects.

表 2-3 2022 年全国因洪涝农作物受灾面积、农作物绝收面积、倒塌房屋情况
Table 2-3 Cropland affected and failed and collapsed dwellings by floods in 2022

地区 Province	农作物受灾面积 / 千公顷 Affected cropland area /1,000 ha	农作物绝收面积 / 千公顷 Failed cropland area /1,000 ha	倒塌房屋 / 万间 Collapsed dwellings/ 10,000 rooms
全国 Nationwide	3413.73	492.65	3.13
北京 Beijing			
天津 Tianjin			
河北 Hebei	25.32	1.94	
山西 Shanxi	142.38	13.82	0.09
内蒙古 Inner Mongolia	497.75	72.11	0.03
辽宁 Liaoning	763.42	149.31	0.02
吉林 Jilin	166.59	16.53	
黑龙江 Heilongjiang	52.93	22.93	
上海 Shanghai			
江苏 Jiangsu			
浙江 Zhejiang	22.06	1.68	0.03
安徽 Anhui	19.06	1.33	
福建 Fujian	60.42	6.52	0.51
江西 Jiangxi	291.87	37.70	0.14
山东 Shandong	22.33	0.58	0.04
河南 Henan	36.24	3.57	0.01
湖北 Hubei	117.80	5.39	0.05
湖南 Hunan	346.16	40.04	0.41
广东 Guangdong	112.90	26.43	0.49
广西 Guangxi	210.77	21.43	0.46
海南 Hainan	0.18	0.03	
重庆 Chongqing	27.01	3.93	0.06
四川 Sichuan	54.46	9.09	0.16
贵州 Guizhou	69.72	9.73	0.07
云南 Yunnan	117.58	15.98	0.04
西藏 Xizang	1.67	0.32	0.01
陕西 Shaanxi	159.33	25.27	0.08
甘肃 Gansu	59.70	4.72	0.39
青海 Qinghai	13.44	1.43	0.02
宁夏 Ningxia	17.82	0.48	0.01
新疆 Xinjiang	4.82	0.36	0.01

注 数据来源于应急管理部，空白表示无灾情。

Note The data come from the Ministry of Emergency Management, and spaces in blank denote no such losses or damages.

图 2-5　2022 年全国洪涝灾害分布图
Figure 2-5　Overview of flood disasters in China in 2022

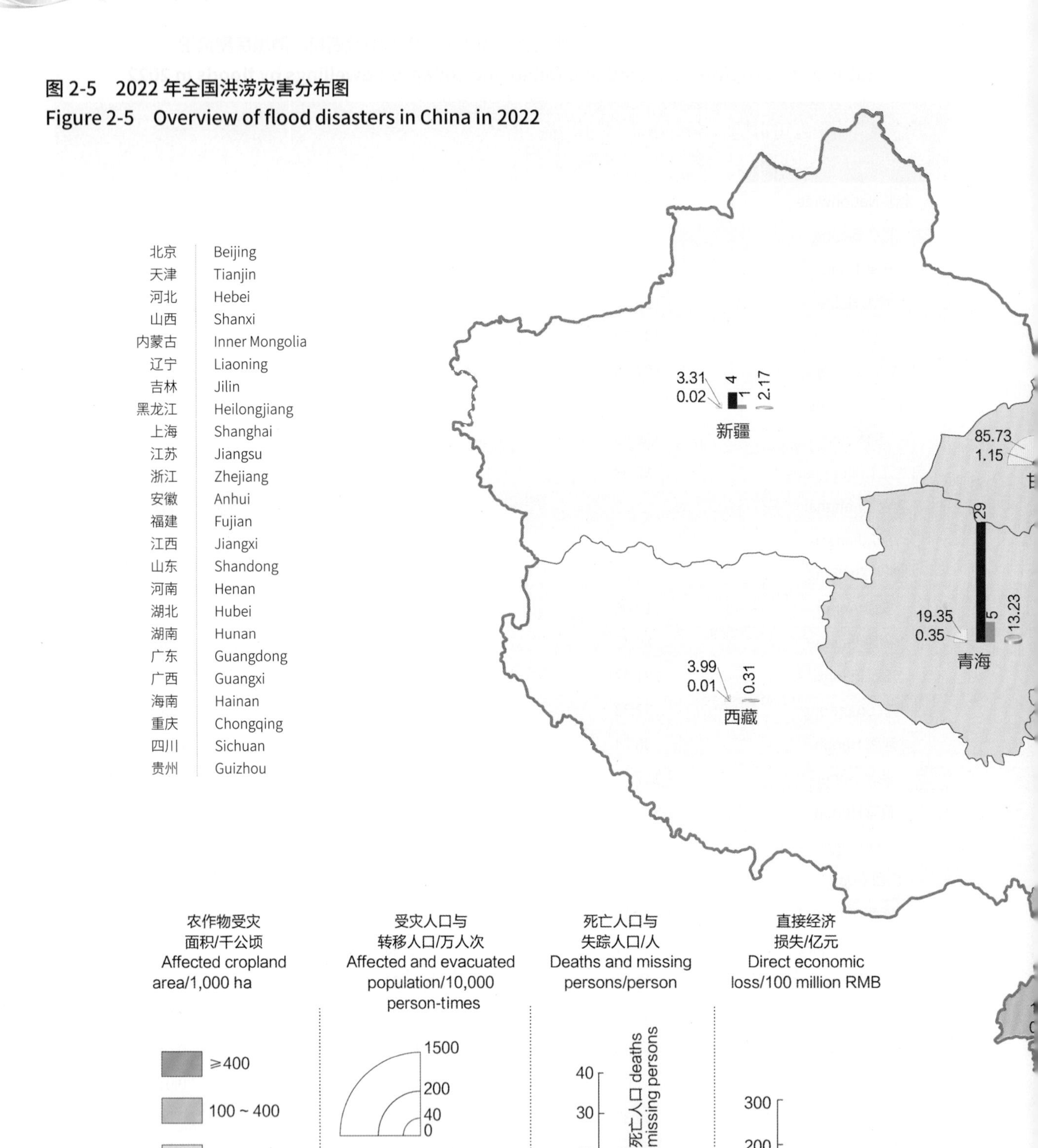

注 数据来源于应急管理部，香港特别行政区、澳门特别行政区、台湾省资料暂缺。

Note The data come from the Ministry of Emergency Management, data of HongKong SAR, Macao SAR and Taiwan are currently unavailable.

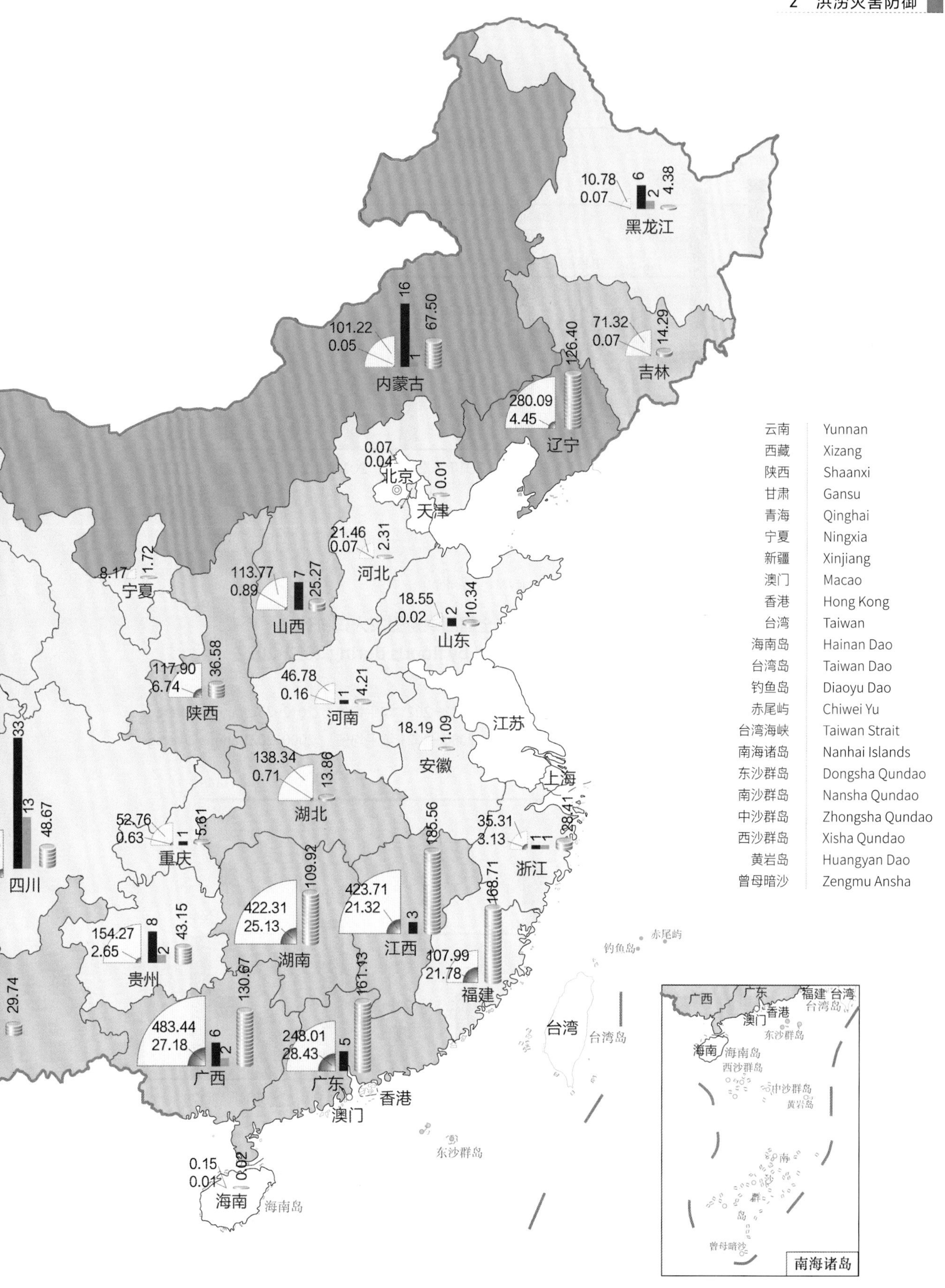

黑龙江
10.78
0.07
6
2
4.38
内蒙古
101.22
0.05
16
1
67.50
吉林
71.32
0.07
14.29
辽宁
280.09
4.45
126.40
北京
0.07
0.04
天津
0.01
河北
21.46
0.07
2.31
宁夏
8.17
1.72
山西
113.77
0.89
7
25.27
山东
18.55
0.02
2
10.34
陕西
117.90
6.74
36.58
河南
46.78
0.16
1
4.21
江苏
安徽
18.19
1.09
上海
湖北
138.34
0.71
13.86
四川
33
13
48.67
重庆
52.76
0.63
1
5.61
浙江
35.31
3.13
1
1
28.41
湖南
422.31
25.13
109.92
江西
423.71
21.32
3
185.56
福建
107.99
21.78
168.71
贵州
154.27
2.65
8
2
43.15
29.74
广西
483.44
27.18
6
2
130.67
广东
248.01
28.43
5
161.13
香港
澳门
海南
0.15
0.01
0.02
海南岛
台湾
台湾岛
钓鱼岛
赤尾屿
东沙群岛
南海诸岛
西沙群岛
中沙群岛
黄岩岛
曾母暗沙
云南 Yunnan
西藏 Xizang
陕西 Shaanxi
甘肃 Gansu
青海 Qinghai
宁夏 Ningxia
新疆 Xinjiang
澳门 Macao
香港 Hong Kong
台湾 Taiwan
海南岛 Hainan Dao
台湾岛 Taiwan Dao
钓鱼岛 Diaoyu Dao
赤尾屿 Chiwei Yu
台湾海峡 Taiwan Strait
南海诸岛 Nanhai Islands
东沙群岛 Dongsha Qundao
南沙群岛 Nansha Qundao
中沙群岛 Zhongsha Qundao
西沙群岛 Xisha Qundao
黄岩岛 Huangyan Dao
曾母暗沙 Zengmu Ansha

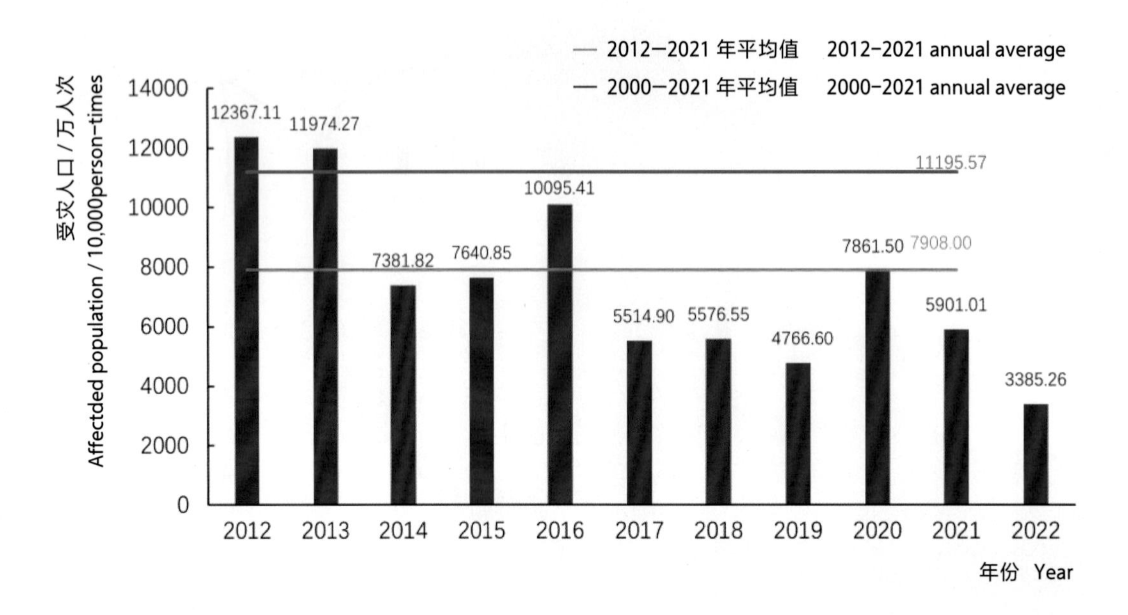

注 2019—2022 年数据来源于应急管理部。

Note The data during 2019–2022 come from the Ministry of Emergency Management.

图 2-6 2012—2022 年全国因洪涝受灾人口统计

Figure 2-6 Population affected by floods during 2012-2022

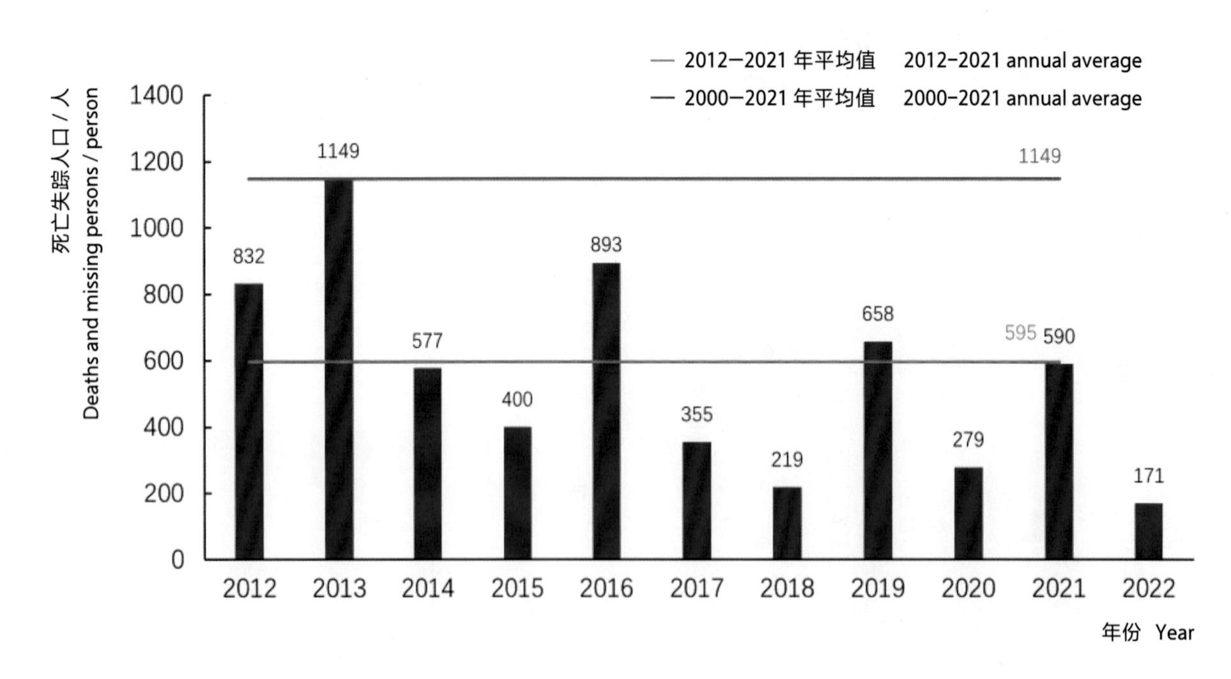

注 2019—2022 年数据来源于应急管理部。

Note The data during 2019–2022 come from the Ministry of Emergency Management.

图 2-7 2012—2022 年全国因洪涝死亡失踪人口统计

Figure 2-7 Deaths and missing persons attributed to floods during 2012-2022

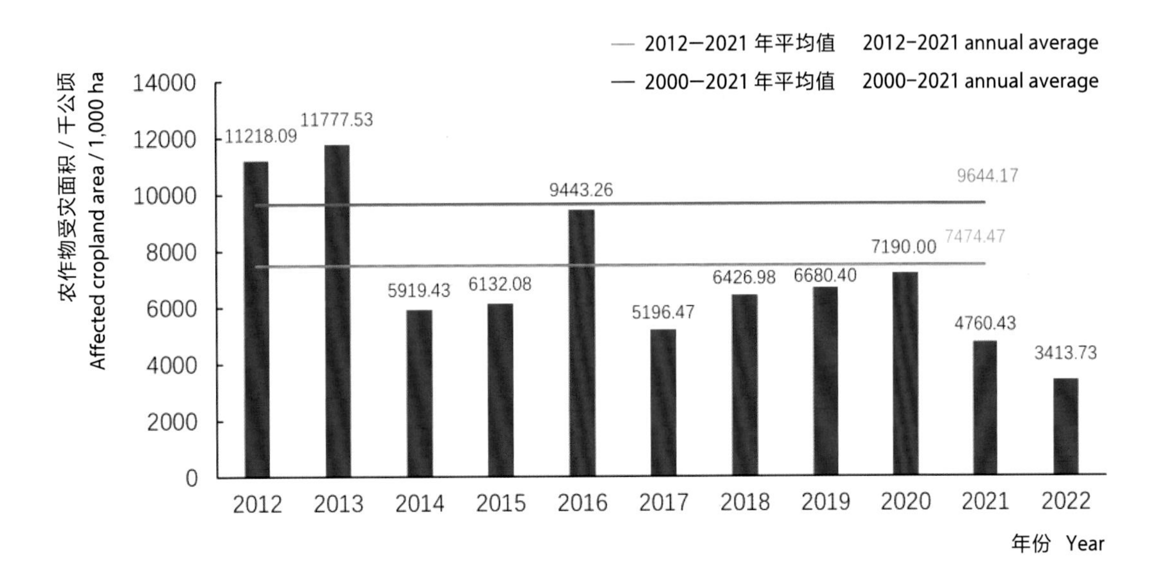

注 2019—2022 年数据来源于应急管理部。

Note The data during 2019–2022 come from the Ministry of Emergency Management.

图 2-8 2012—2022 年全国因洪涝农作物受灾面积统计

Figure 2-8 Cropland area affected by floods during 2012–2022

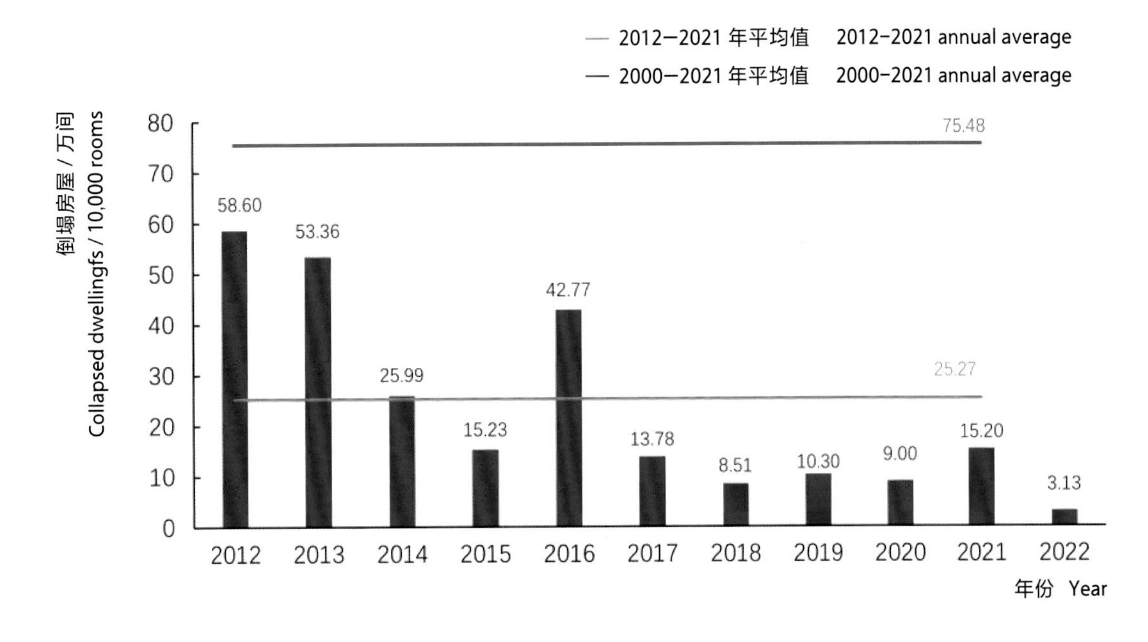

注 2019—2022 年数据来源于应急管理部。

Note The data during 2019–2022 come from the Ministry of Emergency Management.

图 2-9 2012—2022 年全国因洪涝倒塌房屋统计

Figure 2-9 Collapsed dwellings attributed to floods during 2012–2022

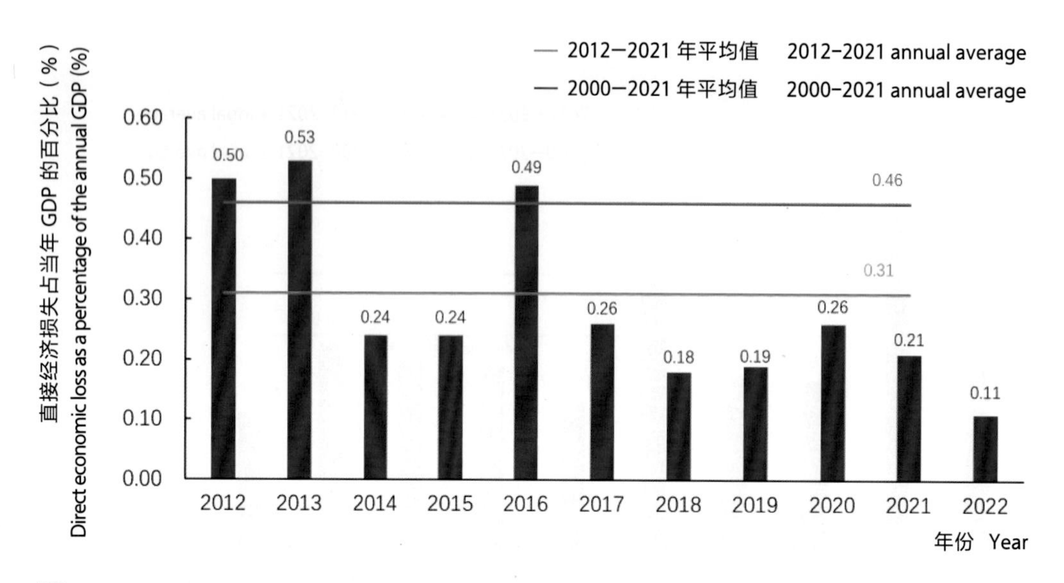

注 2019—2022 年数据来源于应急管理部。

Note The data during 2019-2022 come from the Ministry of Emergency Management.

图 2-10　2012—2022 年全国因洪涝直接经济损失占当年 GDP 的百分比

Figure 2-10　National direct economic losses attributed to floods as a percentage of the annual GDP during 2012-2022

2.4.2　水利工程设施灾损情况

2022 年，全国有 29 省（自治区、直辖市）水利工程设施因洪涝发生损坏，共造成 902 座水库（其中 14 座大型、106 座中型、782 座小型）、26682 处 3606.03 千米堤防（其中 44 处堤防决口）、32496 处护岸、3147 座水闸、10548 座塘坝、55251 处灌溉设施、1938 个水文测站、1918 眼机电井、1753 座机电泵站、774 座水电站（其中 5 座大中型、769 座小型）不同程度受损，水利工程设施直接经济损失 319.12 亿元，较前 10 年平均值下降 25%。其中，水利部直管工程中，长江水利委员会（以下简称长江委）、黄河水利委员会（以下简称黄委）、淮河水利委员会（以下简称淮委）、海河水利委员会（以下简称海委）、珠江水利委员会（以下简称珠江委）、太湖流域管理局（以下简称太湖局）6 个流域管理机构有 2 座水库（1 座大型、1 座小型）、114 处 41.09 千米堤防、440 处护岸、75 个水文测站不同程度受损，水利工程设施直接经济损失 1.57 亿元；水利部松辽水利委员会（以下简称松辽委）水利工程设施未因洪涝受损。

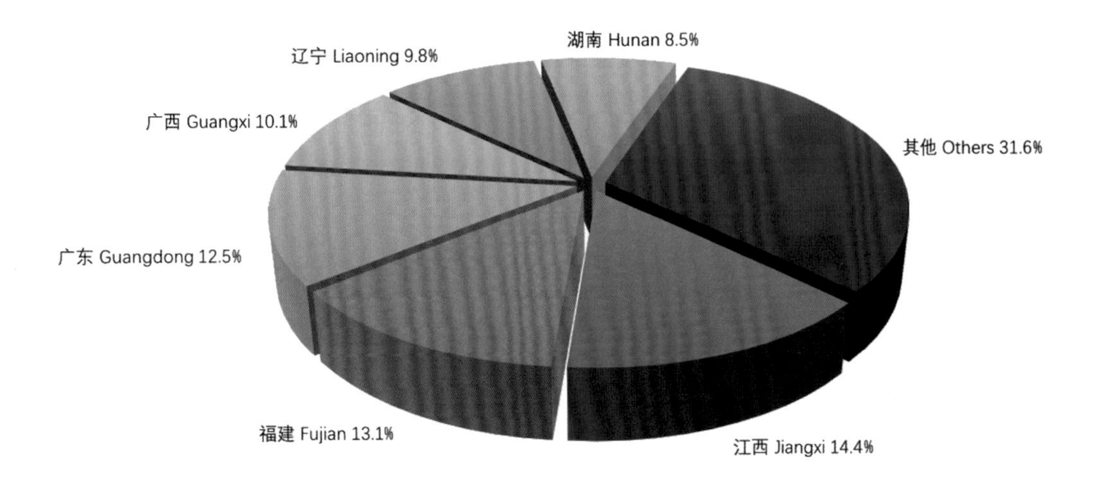

注 数据来源于应急管理部。
Note The data come from the Ministry of Emergency Management.

图 2-11 2022 年全国因洪涝直接经济损失分布
Figure 2-11 A regional break-down of direct economic losses attributed to floods in 2022

2.4.2 Losses and damages to water projects and facilities

In 2022, water projects and facilities in 29 provinces/autonomous regions/municipalities were damaged due to flooding. In total, 902 reservoirs (14 large, 106 medium-sized and 782 small), 26,682 embankments with a total length of 3,606.03 km (including 44 dike breaches), 32,496 bank revetments, 3,147 sluices, 10,548 small pond reservoirs, 55,251 irrigation facilities, 1,938 hydrologic stations, 1,918 electromechanical wells, 1,753 electromechanical pumping stations, and 774 hydropower stations (5 large and medium-sized and 769 small) were damaged to varying degrees, and the direct economic loss billed 31.912 billion RMB, down by 25% from the preceding decadal average. Among the projects directly managed by MWR through the Changjiang Water Resources Commission (hereinafter referred to as the Changjiang Commission), the Yellow River Conservancy Commission (the Yellow River Commission), the Huaihe River Commission (the Huaihe Commission), the Haihe River Water Conservancy Commission (the Haihe Commission), the Pearl River Water Resources Commission (the Pearl River Commission), and The Taihu Basin Authority (the Taihu Authority), 2 reservoirs (1 large and 1 small), 114 embankments totaling 41.09 km in length, 440 bank revetments and 75 hydrologic stations were damaged to varying degrees and the direct economic loss reached 157 million RMB. The projects directly managed by MWR through the the Songliao Water Resources Commission (the Songliao Commission) didn't bear damages due to floods.

表 2-4　2022 年全国水利工程设施灾损情况
Table 2-4　Losses and damages to water projects and facilities in 2022

地区 Province	损坏水库 / 座 Damaged reservoirs/number		损坏堤防 Damaged dikes		损坏护岸 / 处 Damaged revetments/ number	损坏水闸 / 座 Damaged sluices/ number	损坏塘坝 / 座 small Damaged pond reservoirs/ number	损坏水文测站 / 个 Damaged hydrologic stations/ number	损坏水电站 / 座 Damaged hydropower stations/ number	水利工程设施直接经济损失 / 亿元 Direct economic loss by water projects and facilities/100 million RMB
	大中型 Large and medium-size	小型 Small	数量 / 处 Number of sites	长度 / 千米 Length/km						
全国 Nationwide	120	782	26682	3606.03	32496	3147	10548	1938	774	319.12
北京 Beijing										
天津 Tianjin										
河北 Hebei	6	19	678	123.15	514	48	33	63		2.53
山西 Shanxi	5	8	276	100.19	39	2	99	62	6	3.26
内蒙古 Inner Mongolia	7	35	3691	150.61	154	28	23	3		5.90
辽宁 Liaoning	8	22	2059	427.58	2440	124	181	110		19.02
吉林 Jilin	3	9	281	117.56	662	24	50	99	32	6.82
黑龙江 Heilongjiang	1	8	42	10.45	11		3	1		1.09
上海 Shanghai										
江苏 Jiangsu			22	3.70	167	9		9		0.28
浙江 Zhejiang			5112	397.69	2645	143	407	103	59	19.79
安徽 Anhui	1	8	37	7.51	66	19	163	4		0.68
福建 Fujian	2	30	1431	102.52	5057	106	701	236	214	30.70
江西 Jiangxi	3	100	1852	196.07	4568	432	1900	69	120	36.93
山东 Shandong	4	66	714	32.52	202	28	54	186		4.44
河南 Henan	3	17	32	4.19	60	14	10	3		0.65
湖北 Hubei	4	36	411	83.18	482	161	1110	9		8.98
湖南 Hunan	1	17	741	95.81	7767	768	2290	29	14	25.54
广东 Guangdong	15	136	1930	226.98	2262	241	508	28	251	38.90
广西 Guangxi	3	34	1428	189.16	2755	763	407	334	4	27.44
海南 Hainan	4	13	21	0.23	6	10	14	90		1.78
重庆 Chongqing	1	19	494	29.74	550	12	392	81	5	3.97
四川 Sichuan	4	79	1072	155.09	325	89	906	262	15	25.33
贵州 Guizhou	4	9	700	149.93	40	3	40	5	28	5.48
云南 Yunnan		16	462	158.16	113	35	36	7	8	6.10

续表 Continued

地区 Province	损坏水库 / 座 Damaged reservoirs/number		损坏堤防 Damaged dikes		损坏护岸 / 处 Damaged revetments/ number	损坏水闸 / 座 Damaged sluices/ number	损坏塘坝 / 座 small Damaged pond reservoirs/ number	损坏水文测站 / 个 Damaged hydrologic stations/ number	损坏水电站 / 座 Damaged hydropower stations/ number	水利工程设施直接经济损失 / 亿元 Direct economic loss by water projects and facilities/100 million RMB
	大中型 Large and medium-size	小型 Small	数量 / 处 Number of sites	长度 / 千米 Length/km						
西藏 Xizang			275	73.42	30	12	4	46		1.35
陕西 Shaanxi	27	45	1613	209.07	344	21	1043	3	1	15.53
甘肃 Gansu	1	4	675	328.13	211	7	22		1	7.67
青海 Qinghai		4	363	142.87	99	21	49		16	12.75
宁夏 Ningxia	11	47	10	1.20	114		103			0.77
新疆 Xinjiang	1		146	48.23	373	27		21		3.87
长江委直属 Changjiang Water Resources Commission								25		0.02
黄委直属 Yellow River Conservancy Commission		1			436			9		1.21
淮委直属 Huaihe River Commission			1	0.30	2			14		0.06
海委直属 Haihe River Water Conservancy Commission	1		110	40.00	2					0.12
珠江委直属 Pearl River Water Resources Commission								19		0.14
松辽委直属 Songliao Water Resources Commission										
太湖局直属 Taihu Basin Authority			3	0.79				8		0.02

注 空白表示无灾情。

Note Spaces in blank denote no such losses or damages.

2022 年，全国因洪涝造成堤防、护岸、水库、水文站等水利设施损毁主要集中在广东、江西、福建、广西、湖南、四川等 6 省（自治区），水利工程设施直接经济损失占全国的 57.9%。

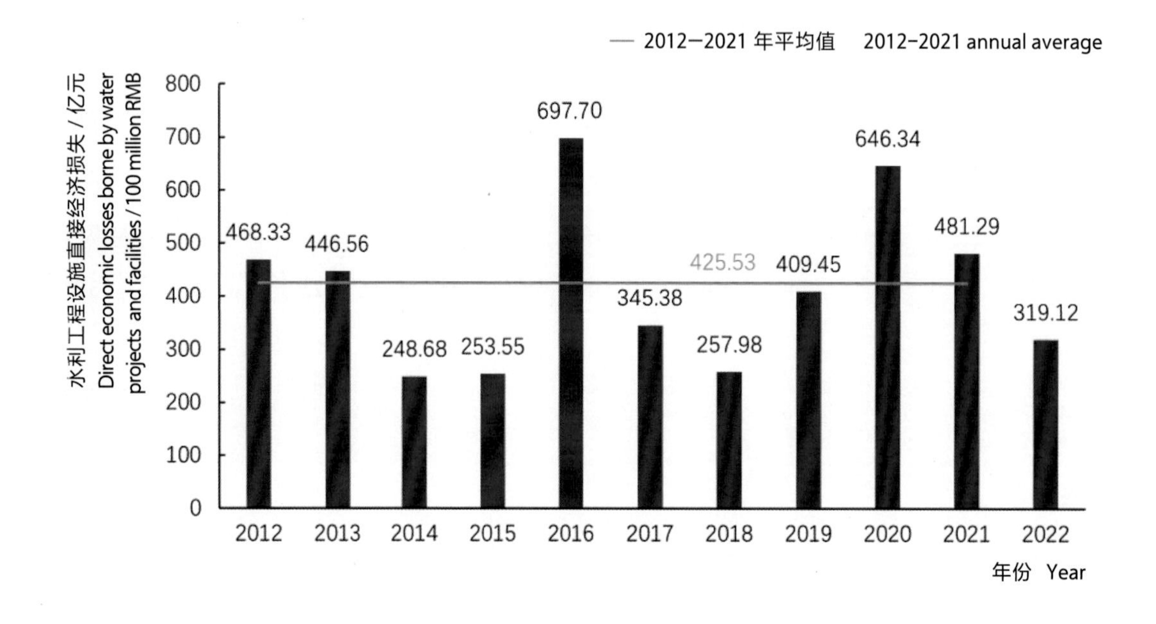

图 2-12 2012—2022 年全国水利工程设施直接经济损失统计
Figure 2-12 Direct economic losses borne by water projects and facilities in China during 2012–2022

In 2022, losses and damages to water facilities such as dikes, revetments, reservoirs, and hydrologic stations attributed to floods nationwide were mainly borne by the six provinces/autonomous regions of Guangdong, Jiangxi, Fujian, Guangxi, Hunan, and Sichuan, and their direct economic losses accounted for 57.9% of the national total of its kind.

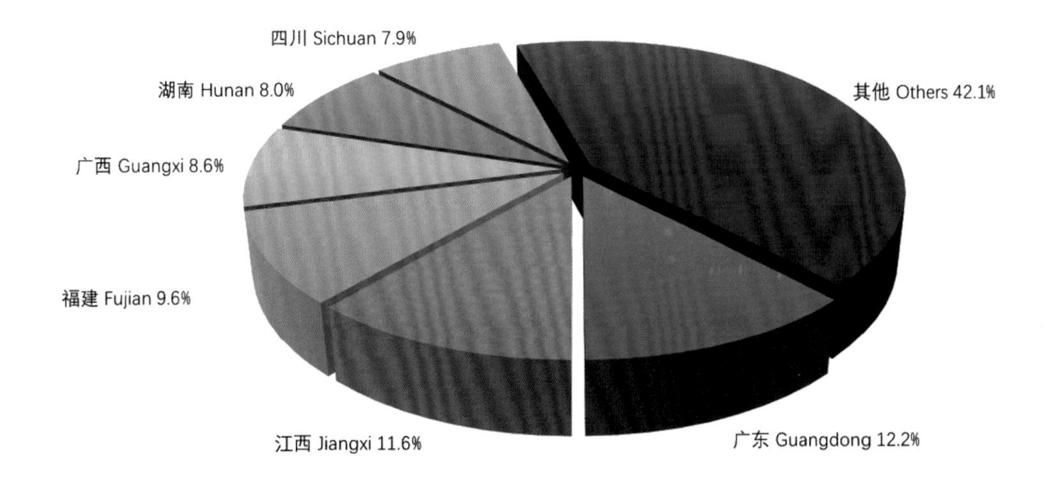

图 2-13　2022 年全国水利工程设施直接经济损失分布

Figure 2-13　A provincial break-down of direct economic losses borne by water projects and facilities in China in 2022

2.5 防御工作

水利部把防汛作为重大政治责任和头等大事，落实预报、预警、预演、预案“四预”措施，贯通雨情、汛情、旱情、灾情“四情”防御，绷紧降雨—产流—汇流—演进、流域—干流—支流—断面、总量—洪峰—过程—调度、技术—料物—队伍—组织“四个链条”，科学精准调度水工程，抓实抓细各项防范应对措施。

2.5.1 工作部署

汛前，水利部召开水旱灾害防御工作视频会议、水库安全度汛视频会议、山洪灾害防御工作视频会议，提早部署防范水旱灾害重大风险，召开专题会议部署重点流域、重点地区防洪保安工作；向社会公布719座大型水库大坝安全责任人名单；针对水库防汛行政、技术、巡查“三个责任人”落实情况和堤防险工险段、穿堤建筑物、水库大坝、溢洪道等关键部位安全隐患进行排查整治；修订印发《水利部水旱灾害防御应急响应工作规程》（水防〔2022〕171号），进一步规范水旱灾害防御应急响应工作程序和应急响应行动，保证水旱灾害防御工作有力有序有效进行，做到防御关口前移；开展七大江河流域防汛备汛线上检查，针对发现的问题，督促地方及时整改；逐流域召开防汛抗旱总指挥部工作会议，开展洪水调度和防御演练，充分发挥流域管理机构组织、指导、协调、监督作用。

2.5 Prevention and Control

MWR regarded flood prevention as a major political responsibility and a top priority. The Ministry implemented the four preemptive pillars of disaster prevention (forecasting, early warning, exercising and contingency planning), integrated prevention and control of rainfall, water regime, hazards, and disasters, strengthened process management from rainfall, runoff generation, confluence to evolution, from the basin scale, mainstreams, tributaries to sections, from flood volume, flood peak, flood processes to scheduling, and from technology, materials, teams to organization, dispatched the water projects in a scientific and accurate manner, and forcefully carried out various preventive and response measures.

2.5.1 Work deployment

Before the flood season, MWR held video conferences on flood and drought disaster prevention, reservoir safety against floods, and flash flood disaster prevention to make early arrangements to prevent major risks of flood and drought disasters, held special meetings to deploy relevant work in key river basins and key areas, announced to the public the list of responsible persons for the safety of the 719 large dammed reservoirs, investigated and rectified the implementation of the "three responsible officials" for reservoir flood control (one administrative official, one technical officer, and one patroller) and the safety hazards of key parts such as risky sections of embankments, buildings crossing the embankments, reservoirs and dams, and spillways, revised and issued the *Regulations of the Ministry of Water Resources for Emergency Response to Flood and Drought Disasters*, further standardized the emergency response procedures and actions, in an effort to ensure that the prevention work was carried out forcefully, orderly and effectively and that the front line of prevention can be as forward as possible. The Ministry also carried out online inspections of flood prevention and preparedness in the seven major river basins, urged local governments to rectify problems in a timely manner, held working meetings of Flood Control and Drought Relief Headquarters (hereinafter FDH) for each river basin, conducted flood dispatch and defense drills, thereby giving full play to the organization, guidance, coordination and supervision roles of river basin authorities/commissions.

2.5.2 隐患排查整改

汛前，重点排查水库、堤防险工险段、病险涵闸和在建水利工程等存在的度汛安全风险隐患，查摆分析水库高水位运用、蓄滞洪区分洪影响因素，完善人员避险转移方案；督促完成水毁修复项目 8821 处，及时恢复水利工程防洪功能；全面清除河道行洪障碍，加强堤防管理和巡查防守，逐一落实穿堤建筑物度汛措施；督促指导地方落实重点流域防洪工程体系运用准备工作；做好水库安全度汛工作，逐库落实防汛“三个责任人”和“三个重点环节”；做好淤地坝防溃坝工作，逐坝落实责任人和抢险措施。

2.5.3 调度演练

水利部根据防洪形势、工程状况和洪水预报，选取历史典型洪水案例，开展洪水调度和防御演练。长江委组织安徽、江西、湖北、湖南、重庆、四川、贵州、云南等省（直辖市）水利厅（局）和长江三峡集团公司等单位开展以 1870 年长江上游特大洪水为背景的防洪调度演练，在汉江丹江口大坝和王甫洲电站开展防汛抢险应急演练。黄委组织开展以郑州“7・20”特大暴雨灾害为背景的 2022 年黄河防御大洪水调度演练，预演典型洪水过程，优化细化洪水调度方案。淮委组织山东省水利厅、江苏省水利厅及沂沭泗水利管理局开展 2022 年沂沭泗河洪水模拟调度推演，提升了防汛调度水平和实战能力。海委联合北京、天津、河北 3 省（直辖市）水利（水务）厅（局）、雄安新区管委会开展大清河洪水防御联合演练，全面检验防汛应急处置能力。珠江委选取珠江“1998・6”历史典型洪水作为背景，考虑降雨量分别增加 10%、15% 等不利情况，首次实现了全流程系统化防洪调度演练。松辽委组织黑龙江省、吉林省、内蒙古自治区水利厅，开展 2022 年松花江流域重要水工程防洪调度演练，实现了预报调度一体化作业和“四预”全过程展示。太湖局组织江苏省水利厅、无锡市水利局等单位以太湖及区域河网处于梅雨期结束后的退水阶段、武澄锡虞区遭遇极端强降雨、预计将对流域和区域防洪安全形成严峻威胁为背景，开展防汛“四预”措施、远程会商、水文应急监测、直管工程运行、险情应急处置等为主要内容的防汛演练。

2.5.2 Investigation and rectification of hazard risks

Before the flood season, efforts were made to investigate the flood hazard risks in reservoirs, dangerous sections of embankments, risky and/or faulty culverts and water projects under construction, analyze the factors affecting the operation of reservoirs at high water levels and flood detention and retention basins, improve the evacuation plans, urge the repair of 8,821 project sites damaged by floods, timely restore the flood control function of water projects, remove obstacles to flood passage, strengthen dike management and patrolling, implement flood control measures for buildings crossing embankments one by one, supervise and guide local governments to prepare for the use of flood control engineering systems in key river basins, implement the “three responsible officials” mechanism and the “three flood control focuses” (forecasting, scheduling, and contingency planning) to ensure reservoir safety during the flood season, and implement the responsible persons and rescue measures for the check dams in order to guard against failure risks.

2.5.3 Dispatching and exercising

Drawing upon the flood development, the projects profiles and the flood forecasts, MWR selected typical historical flood cases to conduct dispatching and flood prevention drills. The Yangtze River Commission organized the water departments (bureaus) of Anhui, Jiangxi, Hubei, Hunan, Chongqing, Sichuan, Guizhou and Yunnan provinces/municipalities and the China Three Gorges Corporation to carry out flood control and dispatch drills using the case of the 1870 extreme floods in the upper Yangtze River, and carried out emergency drills at the Danjiangkou Dam and the Wangfuzhou Power Station on the Hanjiang River. The Yellow River Commission carried out the 2022 Yellow River flood prevention and dispatch drill using the case of the “July 20” extreme rainstorm disaster in Zhengzhou, rehearsed the typical flood process, and optimized the floodwater dispatch plan. The Huaihe Commission organized the Shandong Provincial Water Resources Department, the Jiangsu Provincial Water Resources Department and the Yihe-Shuhe-Sihe Water Management Bureau to carry out the simulation and dispatch of the Yihe-Shuhe-Sihe River flood in 2022, improving on-field floodwater dispatch capabilities. The Haihe Commission, together with the water authorities of Beijing, Tianjin and Hebei Province and the administrative committee of Xiong’an New Area, carried out a joint exercise on flood prevention in Daqing River, comprehensively testing their flood prevention and emergency response capabilities. The Pearl River Commission chose the historical “1998·6” flood in the Pearl River as a typical case, considered even worse conditions such as 10% and 15% increase of rainfall, and completed for the first time a full-process systematic flood control dispatch drill. The Songliao

表 2-5　2022 年汛期各流域管理机构监测预报预警情况

Table 2-5　Monitoring, forecasts and early warnings by each river basin commission/authority during the 2022 flood season

流域管理机构 Commission/authority	降水预报 / 期 Precipitation forecasts/time	洪水预报 / 期 Flood forecasts/time	洪水预警 / 站次 Flood warnings/Station-time	预警短信 / 万条 Alert messages/10,000 pieces
长江委 Changjiang	390	407	12	0.30
黄委 Yellow River	160	281	2	2.30
淮委 Huaihe	140	38	15	0.91
海委 Haihe	122	38		2.20
珠江委 Pearl River		110	63	25.00
松辽委 Songliao	18	72	10	0.97
太湖局 Taihu		167		3.76

注 空白表示无。

Note Blank indicates none.

2.5.4　监测预报预警

按照"降雨—产流—汇流—演进"链条，水利部门加密雨水情监测，发布洪水预报42.4 万站次；按照"流域—干流—支流—断面"链条，加强关键控制性水文断面全要素监控，发布江河洪水预警 2716 次。

汛期，长江委密切关注雨水情变化形势，滚动开展旱涝趋势预测和雨水情预报，加密组织防汛会商，及时发布汛情通报、洪水预警以及短中长期降雨水情预报；针对梅雨期间暴雨洪水过程，会同湖南、江西等省滚动预测预报鄱阳湖、洞庭湖水系雨水情，及时发布预警。黄委坚持主汛期每日会商，遇重要事件加密会商，滚动分析研判雨情、水情、工情，跟踪监测洪水演进，及时发布洪水预警。淮委针对强降雨过程，及时预测降雨分布、降雨总量，加强与流域气象、水文部门沟通，滚动预测预报淮河干流、沂沭泗河及关键性控制断面的洪水总量、洪峰、流量过程，及时发布洪水预警。海委采用精细化智能网格降

Commission organized the water authorities of Heilongjiang Province, Jilin Province and Inner Mongolia Autonomous Region to carry out flood control and dispatch drills for important water projects in the Songhua River basin in 2022, realizing the integrated operation of forecasting and scheduling and the whole process showcase of the four preemptive pillars of flood control and prevention. The Taihu Authority organized the Jiangsu Provincial Water Resources Department and Wuxi Municipal Water Resources Bureau to carry out flood prevention drills focusing on the four preemptive pillars, remote consultation, hydrological emergency monitoring, operation of directly managed projects, and hazards emergency handling. The drill was conducted using the scenario during the receding stage of Taihu Lake and its regional river network after the end of the Meiyu period, when the extremely heavy rain poured in Xiyu District, Wucheng, posed a serious threat to the flood control safety of the river basin and the region.

2.5.4 Monitoring, forecasting and early warning

Bearing in mind the development pattern from rainfall to runoff generation, to confluence, and then to dynamic evolution, the water resources departments intensified the monitoring of rain and water regime conditions and issued flood forecast by 424,000 station-times. According to the chain of "basin scale-mainstream-tributaries-sections", comprehensive monitoring was strengthened for all key hydrological sections, and 2,716 river flood warnings were issued.

During the flood season, the Changjiang Commission closely watched the rainfall development, rolled out trend forecasts and rainfall and flood forecasts, organized frequent consultation meetings, and timely issued flood reports, warnings, and rainfall and flood forecasts in short, middle, and long terms. In view of the heavy rainfall and flood process during the Meiyu period, together with Hunan and Jiangxi provinces, the rainfall and flood development of Poyang Lake and Dongting Lake water system was forecast on a rolling basis, and early warnings were issued in a timely manner. The Yellow River Commission insisted on daily consultations during the main flood season, intensified consultations in case of important events, analyzed the rainfall, floods and project situation on a rolling basis, tracked the evolution of floods and issued flood warnings in a timely manner. In view of heavy rainfall processes, the Huaihe Commission timely predicted the distribution and total amount of rainfall, strengthened communication with the meteorological and hydrological departments of the river basin, and predicted the total flood volume, peak and flow process on the mainstream Huaihe River, the Yihe-Shuhe-Sihe River and key controlling sections on a rolling basis, and issued flood warnings timely. The Haihe Commission adopted refined intelligent grid precipitation forecasting products and flood forecasting coupling technology to improve the 3-day forecast, 7-day forecast and 10-day forecasting mechanism. The

水预报产品与洪水预报耦合技术，完善 3 天预报、7 天预测、10 天展望预报机制。珠江委运用珠江防汛“四预”平台滚动更新预报成果，同步开展洪水监测，“以测补报”提高关键期洪水预报精度，西江、北江等流域重要控制断面的预报误差均在 ±10% 以内。松辽委加强与流域气象中心及流域内 4 省（自治区）水文部门联合会商，开展产汇流分析、洪水演进研究，及时修正洪水预报，关键场次洪水预报精度达到 90% 以上。太湖局构建了流域多目标统筹“四预”一体化系统，持续开展预报模型率定和功能完善，及时发布地区河网水位洪水风险提示、太浦闸倒流关闸预警等。

2.5.5 水工程调度

以流域为单元，遵循系统、统筹、科学、安全原则，综合考虑上下游、干支流、左右岸，精准调度运用水库、河道及堤防、蓄滞洪区为主要组成的流域防洪工程体系，采取“拦、分、蓄、滞、排”措施，累计调度运用大中型水库 4151 座次，拦蓄洪水 925 亿立方米；启用国家蓄滞洪区 1 处，蓄滞洪量 3.1 亿立方米，有效减轻了下游地区防洪压力。

长江委立足防洪保安，兼顾抗旱、防咸潮、供水、灌溉、生态、发电、航运等多目标综合调度需求，实施长江流域 125 座（处）控制性水工程联合调度、统一调度；针对梅雨期鄱阳湖、洞庭湖水系暴雨洪水过程，会同湖南、江西两省滚动预测两湖水系雨水情，科学调度主要水库。黄委加强流域 346 座大型和重点中型水工程防洪调度汛限水位监管，严禁违规超汛限水位运行，压实水库安全监管责任。淮委协调调度出山店水库等水工程拦洪削峰，科学精细调度 12 座直管工程 151 闸次参与泄洪，成功防御 2022 年沭河第 1 号洪水和沂沭泗河多次来水过程，有效防范台风“暹芭”影响。海委统筹防洪与蓄水、防洪与排涝，科学调度骨干水库，最大程度发挥流域水工程体系综合效益。珠江委联合调度西江水系干流龙滩、天生桥一级、岩滩、大藤峡和支流郁江百色、柳江落久、桂江青狮潭等 24 座重点水库，削减梧州站洪峰流量 6000 立方米每秒，降低梧州江段水位 1.80 米，保证了西江沿线防洪安全，减轻了对北江洪水的顶托；指导广东省调度北江水系飞来峡、乐昌峡、湾头等 13 座重点水库拦洪，果断启用潖江蓄滞洪区，及时利用下游的芦苞闸、西南闸分洪，降低北江干流石角站洪峰水位约 0.60 米，控制石角站流量不超过 18500 立方米每秒（北江大堤设计行洪流量 19000 立方米每秒），确保了北江大堤和珠江三角洲防洪安全。松辽委精细调度松花江流域丰满、白山、尼尔基、察尔森等水库，累计拦蓄洪水 88.3 亿立方米，两次协调电力部门减少丰满水库出流，为第二松花江支流鳌龙河洪水下泄创造有利条件，确保第二松花江下游不超警戒；加强对辽河流域水库调度的督导，督促相关省（自治区）预泄腾库、拦洪削峰，滚动开展二龙山、石佛寺、大伙房、观音阁等重点水库来水分析和调洪演算，制定水库调度方案 200 余个，督促吉林省加大二龙山水库出流应对东辽河洪水；协调国家电网加强云峰、水丰等水库调度，应对鸭绿江流域洪水。太

Pearl River Commission used the Pearl River platform of the four preemptive pillars to update the forecast results on a rolling basis, carried out flood monitoring at the same time, and improved the accuracy of flood forecasting in the critical period with "forecasting supplemented by monitoring", and the forecast error of important control sections in the Xijiang and Beijiang river basins is within ±10%. The Songliao Commission strengthened joint consultation with the meteorological center and the hydrological departments of the four provinces and autonomous region in the basin, carried out runoff generation and confluence analysis, flood evolution research, and timely revised flood forecasts, and the accuracy of key flood forecasts exceeded 90%. The Taihu Authority has built an integrated system of the four preemptive pillars for multi-objective overall planning in the river basin, continuously carried out the validation and function improvement of forecast models, and timely issued reminders on water level and flood risk in regional river networks and early warnings for the closure of Taipu Gate because of back flow.

2.5.5 Water project scheduling

Scheduling of the flood control engineering projects–reservoirs, river channels, embankments, and flood detention and retention basins–was conducted from the whole basin perspective and in the principles of being "systematic, coordinated, scientific and safe"; considerations were given to both the upstream and the downstream, mainstream and tributaries, and the left and right banks; measures including "blocking, dividing, detention, retention and discharging" were used. Cumulatively, the large and medium-sized reservoirs were dispatched for 4,151 times, detaining and holding 92.5 billion m^3 of floodwater. One national flood detention and retention basin was mobilized to hold 310 million m^3 of floodwater, effectively reducing the flood control pressure downstream.

Focusing on flood control while also fulfilling the multi-functional needs of drought relief, salt tide control, water supply, irrigation, ecology, power generation, and navigation, the Changjiang Commission implemented joint and unified dispatch of 125 controlling water projects in the Yangtze River basin. To handle the rainstorm and flood process in Poyang Lake and Dongting Lake during the Meiyu period, the Commission worked with Hunan and Jiangxi provinces to predict the rainfall and flood conditions of the two lakes on a rolling basis and scientifically dispatched the key reservoirs. The Yellow River Commission strengthened the supervision of flood control level compliance in the 346 large and key medium-sized water projects, strictly prohibited the illegal operation of exceeding the flood control level, and emphasized the responsibility for the safety supervision of reservoirs. The Huaihe Commission coordinated and dispatched water projects such as the Chushandian Reservoir to retain floods, cut and delay peaks, mobilized the 12 directly managed projects to participate in flood discharge for 151 gate-times, successfully controlling the No. 1 Flood of the Shuhe River in 2022 and the multiple rounds of floodwater rise in the Yihe-Shuhe-Sihe

湖局在台风“梅花”影响期间，充分发挥太湖调蓄作用，关闭环太湖“内圈”口门，减轻下游杭嘉湖、上海等地防洪压力，及时调度常熟水利枢纽全力排水，同时督促沿长江、沿杭州湾、沿海等“外圈”口门加大排水力度，累计泄水 2.7 亿立方米，有效降低了地区河网水位。

River, and effectively preventing the impact of Typhoon "Chaba". The Haihe Commission coordinated the relationship between flood control and water storage, and also between flood control and drainage; scientifically dispatched the backbone reservoirs and maximized the comprehensive benefits of the river basin water engineering system. The Pearl River Commission jointly dispatched 24 key reservoirs, including Longtan, Tianshengqiao I, Yantan, and Datengxia on the mainstream, and those on the tributaries, such as Baise on Yujiang River, Luojiu on Liujiang River and Qingshitan on Guijiang River, to reduce the peak flow at Wuzhou Station by 6,000 m^3/s and draw down the water level of the Wuzhou River section by about 1.80 m, ensuring the safety of flood control along the Xijiang River and facilitating the floodwater of the Beijiang River to pass. The Commission also guided Guangdong Province to dispatch 13 key reservoirs in the Beijiang River system, including Feilaixia, Lechangxia and Wantou, decisively used the Pajiang River flood detention and retention basin, and made timely use of the Lubao Gate and Xinan Gate downstream to divert floodwater, thereby reducing the peak water level at Shijiao Station on the mainstream Beijiang by about 0.60 m and controlling the flow of Shijiao Station to not exceed 18,500 m^3/s (the design flood flow of the Beijiang embankment being 19,000 m^3/s). The flood control safety of the Beijiang Embankment and the Pearl River Delta was guaranteed. The Songliao Commission dispatched the Fengman, Baishan, Nierji and Chaersen reservoirs in the Songhua River basin to retain and hold 8.83 billion m^3 of floodwater; the Commission twice coordinated with the power authorities to reduce the outflow of the Fengman Reservoir to create favorable conditions for the discharge of the Aolong River, a tributary of the Second Songhua River. This ensured that the lower reaches of the Second Songhua River did not exceed the warning water level. The Commission strengthened its supervision on the reservoir dispatch in the Liaohe River basin by urging the related provinces and autonomous regions to draw down the reservoir level in preparation for flood control, and retain floods and cut peaks, carried out inflow analysis and flood regulation calculation on a rolling basis for key reservoirs such as Erlongshan, Shifosi, Dahuofang and Guanyinge, formulated more than 200 reservoir dispatch plans, urged Jilin Province to increase the outflow of the Erlongshan Reservoir to cope with the flooding of the Dongliao River, and coordinated with the State Grid to strengthen the dispatch of Yunfeng and Shuifeng reservoirs to cope with floods in the Yalu River basin. During the impact of Typhoon "Muifa", the Taihu Authority gave full play to the role of Taihu Lake in regulating and storing water by closing the entrance gate of the "inner circle" around Taihu Lake, reducing the flood control pressure on Hangjia Lake and Shanghai downstream, timely dispatched the Changshu water conservancy complex to drain water, and urged the "outer circle" gates along the Yangtze River, around the Hangzhou Bay, and on the coast to increase drainage. With a total discharge of 270 million m^3, the water level of the regional river network was effectively reduced.

2.5.6 会商响应

水利部坚持汛期24小时值守，建立健全主汛期逐日会商机制和部领导周专题会商机制，滚动会商研判186次，会商意见直达相关地区防御一线，每天以"一省一单"形式将预报降雨量超过预警阈值（50毫米或25毫米）的县（市、区）和水库名单发至相关地方，提醒做好强降雨防范；启动洪水灾害防御应急响应16次，指导流域管理机构启动洪水灾害防御应急响应59次；水利部及各流域管理机构共派出234个次工作组、专家组，协助指导地方开展防御工作。

2.5.6 Consultation and responses

MWR kept on 24-hour duty during the flood season, established and improved the mechanism of daily consultation during the main flood period and the special weekly consultation mechanism by the ministerial leadership, conducted 186 dynamic consultations to share the results directly with the front line of flood control in relevant areas, and sent the list of counties/cities/districts and reservoirs where the forecast rainfall would exceed the warning threshold (50 mm or 25 mm) to relevant places on a daily basis in the form of "one province, one report" The Ministry launched 16 emergency responses for flood disaster prevention, and guided river basin commissions/authorities to initiate such emergency responses for 59 times. The Ministry and river basin commissions/authorities dispatched a total of 234 working groups and expert teams to assist and guide local flood disaster prevention work.

表 2-6 2022年水利部本级洪水灾害防御应急响应启动情况
Table 2-6 Initiation of flood disaster prevention emergency response by the Ministry of Water Resources in 2022

序号 No.	启动日期 Start Date	响应级别 Response level	事由 Cause	终止日期 End Date
1	5月10日 May 10	IV	华南暴雨洪水过程，部分中小河流发生超警戒以上洪水，北江发生编号洪水 During the rainstorm and flood process in South China, some small and medium-sized rivers experienced floods above the warning level, and numbered floods occurred in the Beijiang River	5月15日 May 15
2	5月27日 May 27	IV	云南、广西部分地区连续降雨，部分地区中小河流发生洪水，山洪灾害防御存在较大风险 Continuous rainfall in parts of the Yunnan and Guangxi led to high risks of flooding and flash flood in the medium and small rivers in some areas	5月31日 May 31
3	5月28日 May 28	IV	浙江、安徽、福建、江西、湖南、广东、贵州等地发生强降雨过程 Heavy rainfall processes in Zhejiang, Anhui, Fujian, Jiangxi, Hunan, Guangdong and Guizhou	5月31日 May 31

续表 Continued

序号 No.	启动日期 Start Date	响应级别 Response level	事由 Cause	终止日期 End Date
4	6月3日 June 3	IV	江西、湖南、广西、贵州等地发生较大洪水 Medium floods in Jiangxi, Hunan, Guangxi and Guizhou	6月10日 June 10
5	6月6日 June 6	IV	浙江、福建、广东、云南等地发生强降雨过程 Heavy rainfall processes in Zhejiang, Fujian, Guangdong and Yunnan	6月10日 June 10
6	6月12日 June 12	IV	福建、江西、湖南、广东、广西、贵州、云南等地发生强降雨过程 Heavy rainfall processes in Fujian, Jiangxi, Hunan, Guangdong, Guangxi, Guizhou and Yunnan	6月23日 June 23
7	6月13日 June 13	III	珠江流域将发生流域性较大洪水 The Pearl River basin is bracing for basin-wide medium flood	6月25日 June 25
8	6月18日 June 18	IV	浙江、安徽等地发生强降雨过程 Heavy rainfall process in Zhejiang and Anhui	6月23日 June 23
9	6月30日 June 30	IV	广东、广西、海南等地预报将受台风影响 Guangdong, Guangxi and Hainan to be affected by typhoon according to forecasts	7月9日 July 9
10	7月10日 July 10	IV	山西、陕西、甘肃、河北、河南等地发生强降雨过程 Heavy rainfall processes in Shanxi, Shaanxi, Gansu, Hebei and Henan	7月20日 July 20
11	7月17日 July 17	IV	辽河发生2022年第1号洪水 No.1 Floood of Liaohe River in 2022	8月17日 August 17
12	8月6日 August 6	IV	北京、天津、河北、山西、内蒙古、山东、河南、陕西等地发生强降雨过程，暴雨区部分河流可能发生超警戒洪水 Heavy rainfall processes in Beijing, Tianjin, Hebei, Shanxi, Inner Mongolia, Shandong, Henan and Shaanxi; potentially floods above the warning level might hit some rivers in the rainstorm zone	8月17日 August 17
13	8月9日 August 9	IV	广东、广西、海南、云南、贵州等地发生强降雨过程，暴雨区部分中小河流可能发生超警戒洪水 Heavy rainfall processes in Guangdong, Guangxi, Hainan, Yunnan and Guizhou; potentially floods above the warning level might hit some medium to small rivers in the rainstorm zone	8月13日 August 13
14	8月23日 August 23	IV	广东、广西、海南、贵州、云南等地受2209号台风“马鞍”影响，发生强降雨过程 Heavy rainfall processes in Guangdong, Guangxi, Hainan, Guizhou and Yunnan due to Typhoon “Ma-on” (No. 2209)	8月27日 August 27
15	9月6日 September 6	IV	四川甘孜州泸定县发生6.8级地震 Magnitude 6.8 earthquake in Luding County, Ganzi Prefecture, Sichuan Province	9月10日 September 10
16	9月12日 September 12	IV	上海、浙江、江苏、福建、山东等地受2212号台风“梅花”影响，发生强降雨过程 Heavy rainfall processes in Shanghai, Zhejiang, Jiangsu, Fujian and Shandong due to Typhoon “Muifa” (No. 2212)	9月17日 September 17

表 2-7　2022 年各流域管理机构会商及洪水灾害防御应急响应启动情况
Table 2-7　Consultations held and emergency respon ses initiated by river basin commissions / authorities for flood disaster prevention in 2022

流域管理机构 Commissions / authorities	会商次数 Consultations	启动次数 Emergency responses				累计时间 / 天 Accumulation time / days
		Ⅰ级 Level Ⅰ	Ⅱ级 Level Ⅱ	Ⅲ级 Level Ⅲ	Ⅳ级 Level Ⅳ	
长江委 Changjiang	115				4	24
黄委 Yellow River	66				2	45
淮委 Huaihe	116			1	7	18
海委 Haihe	27			1	7	14
珠江委 Pearl River	138	1	1	8	13	42
松辽委 Songliao	58			2	4	53
太湖委 Taihu	127		1	2	5	38
合计 Total	647	1	2	14	42	234

注 空白表示未启动。
Note Blank means zero initiation.

2.5.7 部门协作

水利部强化与应急管理部联合会商，共享预测预报情况及汛情灾情信息；会同国家发展改革委、财政部、住房城乡建设部稳步推进防汛抗旱水利提升工程建设；商财政部安排水利救灾资金 16.96 亿元，支持地方做好安全度汛隐患整治、防洪工程设施灾损修复工作；完善与气象部门山洪灾害气象预警联合发布机制；派出水利专家配合有关部门赴巴基斯坦完成防洪减灾指导等任务。各流域管理机构坚持流域统筹，做好信息共享、方案协调、统一调度等工作。地方各级水利部门上下联动、密切配合，细化实化各项防御措施，形成了洪涝灾害防御强大合力。

辽宁绕阳河盘锦曙四联段溃口处堤坝成功合龙（8 月 6 日）
The embankment breach in Shusilian section on the Raoyang River was successfully closed, Panjin, Liaoning Province (August 6)

2.5.7 Cross-sectoral collaboration

MWR strengthened joint consultations with the Ministry of Emergency Management to share forecasts and flood disaster information; worked with the National Development and Reform Commission, the Ministry of Finance and the Ministry of Housing and Urban-Rural Development to advance the construction of upgrading projects for flood control and drought mitigation; worked with the Ministry of Finance to arrange 1.696 billion RMB of water-related disaster relief funds to support local governments in rectifying potential flood safety hazards and repairing flood control facilities; improved the joint mechanism with meteorological departments for issuing meteorological warnings for flash flood disaster; sent water conservancy experts to Pakistan to cooperate with relevant departments to complete tasks such as flood control and disaster reduction guidance. All river basin commissions/authorities adhered to the overall planning of river basins, and implemented information sharing, program coordination, and unified scheduling, etc. Local water agencies at all levels worked closely together to refine and implement various defensive measures, forming a strong joint force for flood disaster prevention.

2.5.8 支撑保障

充分发挥水利防洪抢险技术支撑作用。水利部共派出 66 个工作组、专家组深入防御一线指导地方做好强降雨防范和各类险情处置，协助成功处置广东曲江区苍村水库溢洪道险情、辽宁绕阳河曙四联段溃口险情、内蒙古通辽市科左后旗东五家子水库险情等。全国水利系统共派出工作组 7.16 万组次 31.96 万人次、专家组 3.14 万组次 13.8 万人次赴一线指导开展巡查防守、险情处置工作。

2.6 防御成效

2022 年，水利部门科学调度运用水库、河道及堤防、蓄滞洪区，有效应对了江河洪水，全国水库无一垮坝，大江大河干流堤防无一决口，全年因洪涝死亡失踪人口为 1949 年以来最少。全国减淹城镇 1649 座次，减淹耕地 1020.23 千公顷，避免人员转移 690.26 万人次，最大程度保障了人民群众生命财产安全及重要基础设施安全运行。

表 2-8 2022 年各流域减淹城镇、耕地及避免人员转移情况
Table 2-8 Cities/towns and cropland protected from floods, and population avoiding evacuation in 2022

流域 River/lake basin	减淹城镇 / 座次 Cities/towns cumulatively protected from floods	减淹耕地 / 千公顷 Cropland protected from floods/1,000 hectares	避免人员转移 / 万人次 Population avoiding evacuation/10,000 person-times
全国 Nationwide	1649	1020.23	690.26
长江流域 The Yangtze River	328	172.05	131.27
黄河流域 The Yellow River	55	51.55	30.76
淮河流域 The Huaihe River	159	46.89	6.16
海河流域 The Haihe River	18	2.33	2.53
珠江流域 The Pearl River	589	402.75	352.49
松辽流域 The Songhua-Liaohe River	382	284.77	100.24
太湖流域 The Taihu Lakes	118	59.89	66.81

2.5.8 Support and guarantee

The technical support role of water agencies in flood control and rescue was given full play. MWR dispatched a total of 66 working groups and expert teams to the front line to guide local governments in preventing disasters caused by heavy rainfall and dealing with various hazards, assisting in successfully handling the spillway hazard of the Cangcun Reservoir in Qujiang District, Guangdong Province, the collapse of the Shusilian Section of the Raoyang River in Liaoning, and the Dongwujiazi Reservoir hazard in Kezuohou Banner of Tongliao City, Inner Mongolia. The national water agencies sent out 71,600 group-times of 319,600 person-times in working groups, and 31,400 team-times of 138,000 person-times in expert teams to the front line to guide the work to conduct patrol and defense and handle various hazards.

2.6 Effectiveness of Flood Disaster Prevention

In 2022, the water agencies scientifically dispatched reservoirs, river channels and embankments, and flood detention and retention basins to effectively cope with river floods, and none of the reservoirs nation-wide collapsed, and none of the main embankments of major rivers were broken, and the number of deaths and missing persons attributed to floods throughout the year was the lowest since 1949. Cumulatively, 1,649 cities and towns and 1,020,230 ha of cropland were protected from being flooded nationwide, while the evacuation of 6,902,600 person-times were avoided, ensuring the safety of lives and property and the safe operation of important infrastructure to the greatest extent.

案例 1 珠江流域洪水调度

5 月下旬至 7 月上旬，受连续强降雨和台风“暹芭”影响，珠江流域西江发生 4 次编号洪水、北江发生 3 次编号洪水、韩江发生 1 次编号洪水，其中北江第 2 号洪水为 1915 年以来最大洪水。

防御西江第 4 号洪水和同期北江第 2 号洪水期间，珠江委统筹流域全局，强化流域统一调度，共发出 30 余道调度令，会同广西壮族自治区水利厅调度西江干支流 24 座水库，共拦蓄洪水 38 亿立方米，其中，调度龙滩、百色等骨干水库全力拦洪，调度红水河岩滩、柳江落久、郁江西津、桂江青狮潭等支流水库群适时错峰，调度在建大藤峡水库工程发挥关键作用精准削峰，削减梧州洪峰流量 6000 立方米每秒，降低梧州洪峰水位 1.80 米，将西江洪峰延后 38 小时，避免西江、北江洪水恶劣遭遇，为北江洪水安全宣泄创造了有利条件，极大地减轻了西江下游和珠江三角洲的防洪压力。与此同时，珠江委及时向广东省水利厅提出北江蓄洪、滞洪、分洪调度建议方案，指导启用潖江蓄滞洪区，先后启用了独树围、踵头围、大厂围、江咀围、下岳围共 5 个堤围滞洪 3.1 亿立方米，转移区内人口近 4 万人；会同广东省水利厅精细化调度乐昌峡、湾头、锦江、孟洲坝、白石窑、飞来峡等 13 座重点水库拦洪削峰，及时利用下游的芦苞闸、西南闸分洪，将飞来峡入库流量 19900 立方米每秒削减至石角站洪峰流量 18500 立方米每秒，洪水量级控制在北江大堤安全泄量以内，确保了北江大堤和珠江三角洲城市群防洪安全。

Case 1 Flood control and dispatch in the Pearl River basin

From late May to early July, due to continuous heavy rainfall and the impact of typhoon "Chaba", four numbered floods occurred in the Xijiang River, three numbered floods in the Beijiang, and one numbered flood in the Hanjiang in the Pearl River basin. In particular, the Beijiang River No. 2 Flood was its largest flood since 1915.

In fighting against the No. 4 Flood of the Xijiang River and the concurrent No. 2 Flood of the Beijiang River, the Pearl River Commission had coordinated and strengthened the unified dispatch within the basin, issued more than 30 dispatch orders, and scheduled 24 reservoirs on the mainstream Xijiang and its tributaries in conjunction with the water resources department of Guangxi Zhuang Autonomous Region. A total of 3.8 billion m^3 of floodwater was detained and stored. The backbone reservoirs such as Longtan and Baise were dispatched to hold back floods as much as possible; the reservoir groups on tributaries such as Yantan on the Hongshui River, Luojiu on the Liujiang River, Xijin on the Yujiang River, and Qingshitan on the Guijiang River were dispatched to delay the flood peak in a timely manner; and the dispatch of the Datengxia Reservoir Project, still under construction, played a key role in precisely cutting the peaks, thereby reducing the peak flow of Wuzhou station by 6,000 m^3/s, lowering its peak water level by 1.80 m, and delaying the peak of the Xijiang River by 38 hours. As a result, the convergence of floods from the Xijiang and Beijiang rivers was mitigated, creating favorable conditions for the safe flood discharge of the Beijiang River and greatly alleviating the flood control pressure on the lower Xijiang River and the Pearl River Delta. At the same time, the Pearl River Commission promptly put forward to the Guangdong Provincial Water Resources Department a proposal for the dispatch of flood storage, detention and diversion in the Beijiang River, guided the use of the Pajiang flood detention and retention basin, and successively activated five embankments to detain a total of 310 million m^3 of floodwater, including Dushuwei, Zhongtouwei, Dachangwei, Jiangzuiwei and Xiayuewei. Nearly 40,000 people in the area were relocated. In cooperation with the Guangdong Provincial Water Resources Department, the Commission precisely dispatched 13 key reservoirs, including Lechangxia, Wantou, Jinjiang, Mengzhouba, Baishiyao and Feilaixia, for flood detention and peak cutting, and used the Lubao and Xinan gates downstream to divert floodwater, successfully reducing the inflow at Feilaixia (19,900 m^3/s) to 18,500 m^3/s at Shijiao Station. The flood was controlled within the safe discharge volume of the Beijiang embankment, ensuring the flood control safety of the Beijiang embankment and the city groups in the Pearl River Delta.

案例 2 辽河支流绕阳河曙四联段溃口处置

2022 年 8 月 1 日 10 时 30 分，辽宁绕阳河左岸堤坝曙四联段发生溃口。溃口险情发生后，盘锦市及时启动防汛Ⅰ级应急响应，并宣布溃口影响区域进入紧急防汛期。水利部及松辽委立即成立工作组，赶赴溃口现场协助开展抗洪抢险工作，为地方提供抢险技术支撑。

工作组首先要求地方做好人员转移安置工作，确保应转尽转、不落一人，切实保障受灾群众基本生活，严防灾害风险解除前人员擅自返回造成伤亡；建议地方尽快分析确定溃口后对淹没区域的影响范围及水深，确定第三道防线的抢筑高程，加快构建以省道 308、省道 102 和曙十三支等组成的第三道防线，并采取防风、防雨、防浪措施，做好第三道防线的防守，严防淹没区蔓延；建议地方加密溃口处水情监测，尽快调配溃口封堵所需的材料、设备、人员，应急调度红旗水库，在确保安全的前提下，将 44 孔闸门下落，最大限度控制出库流量，达到入库流量与下泄流量持平，减轻了绕阳河下游行洪压力，为溃口封堵创造有利条件。经多方共同努力，8 月 6 日 18 时 20 分成功封堵溃口。

溃口处完成封堵后，水利部督促地方尽快开展溃口段堤防加高培厚、防渗闭气等封堵后续工作，并做好淹没区洪水强排、污染物清理、卫生防疫和退水期、排水期堤防的巡查防守工作；指导辽宁省水利厅、辽河油田等单位调集 25 支排水队 1145 人、220 台大型排水设备紧急开展应急排水，累计排水 8839 万立方米，其中受淹区排水 7804 万立方米、油田作业区排水 1035 万立方米。

Case 2 Response to the dyke breach at the Shusilian section of the Raoyang River, a tributary of the Liaohe River

At 10:30 on August 1, 2022, a breach occurred in the Shusilian dyke section on the left bank of the Raoyang River in Liaoning. After the breach occurred, Panjin City promptly activated Level I emergency response for flood control and declared emergency state for the affected area. A working group swiftly set up by MWR and the Songliao Commission arrived at the breach site to assist in flood control and rescue and provide local emergency technical support.

The working group first required early and complete evacuation and relocation to make sure that no affected person be left behind and that the basic subsistence needs of the people affected be guaranteed. Efforts were made to prevent casualties caused by unauthorized return before risks were lifted. The working group advised that the scope of impact and the water depth of the flooded area after the breach be identified as soon as possible so as to determine how high the emergency third defense line should be built; the construction of the third defense line composed of provincial highway No.308, provincial highway No.102 and Shushisanzhi Road was accelerated, and measures were taken to guard against wind, rain and waves and to prevent the flooded area from spreading beyond the third defense line. It was advised that the flooding situation at the breach be closed monitored, and materials, equipment and personnel required for breach blocking be ready as soon as possible. Under safe operation, the 44-hole gate of the Hongqi Reservoir was closed to bring down outflow on par with the inflow, in an effort to reduce the flood pressure on the lower Raoyang River and allow for breach blocking. The breach was blocked at 18:20 on August 6 under such joint efforts.

After the breach was blocked, MWR urged for swift wrapping up of the remaining work, including embankment heightening and thickening and seepage prevention. In addition, efforts were also made to artificial floodwater drainage in the inundated area, pollutant clearing, sanitation and epidemic prevention, and reinforced embankment inspection and patrolling during the flood receding and draining period. Guidance was provided to agencies including the Liaoning Provincial Water Resources Department and Liaohe Oilfield (under CNPC) to mobilize 25 drainage teams of 1,145 personnel and 220 large drainage equipment to carry out emergency drainage. A total of 88.39 million m^3 of floodwater, including 78.04 million m^3 in flooded areas and 10.35 million m^3 in oilfield operation areas, were drained.

3 山洪灾害防御

FLASH FLOOD DISASTER PREVENTION

3.1 基本情况

2022 年，全国共发生 17 起有人员死亡失踪的山洪（山洪泥石流）灾害事件，因灾死亡失踪 119 人（其中因山洪灾害死亡失踪 74 人，因山洪泥石流灾害死亡失踪 45 人），占全国因洪涝灾害死亡失踪人口的 69.6%，较 2011—2021 年因灾平均死亡失踪人口下降 65.5%，为有统计资料以来最少。

3.1 Disasters and Losses

In 2022, China witnessed 17 flash flood (flash flood and debris flow) incidents that had led to life losses: 119 people were killed or went missing (including 74 deaths and 45 missing persons attributed to flash flood disasters), accounting for 69.6% of the deaths and missing persons attributed to all flood disasters. It registered a 65.5% drop from the 2011–2021 annual average in the same category, hitting the lowest since records began.

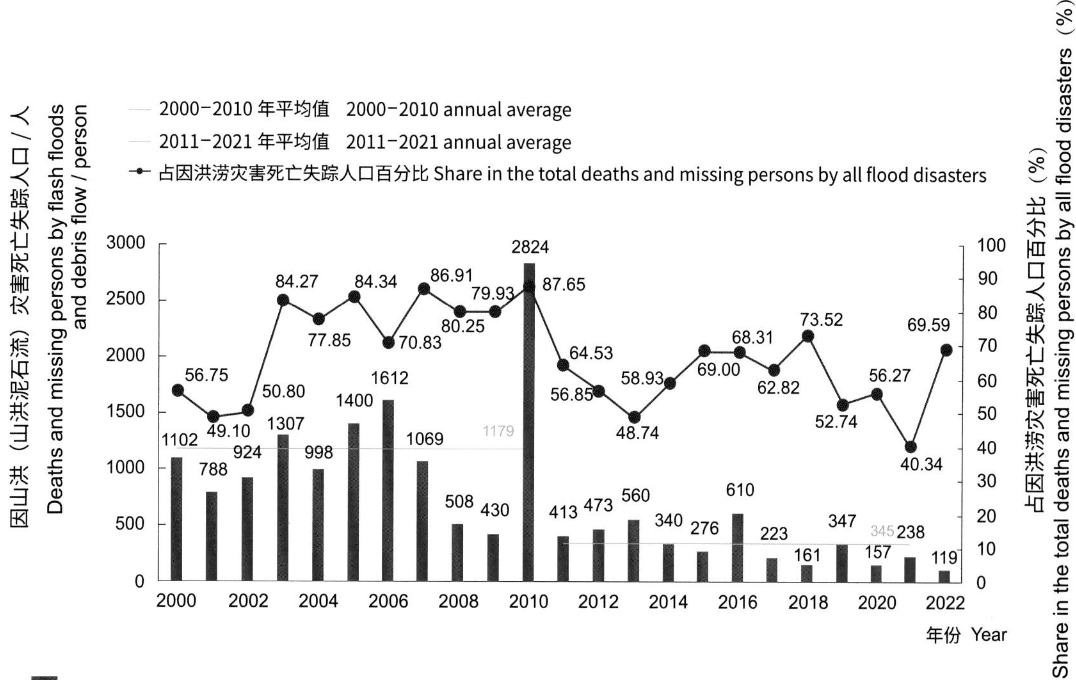

注 2000—2010 年数据为因灾死亡人口；2011—2022 年数据为因灾死亡失踪人口。

Note Statistics during 2000–2010 include only deaths attributed to flash floods; statistics during 2011–2022 include both deaths and missing persons attributed to flash floods.

图 3-1 2000—2022 年因山洪（山洪泥石流）灾害死亡失踪人口及占因洪涝灾害死亡失踪人口的百分比

Figure 3-1 Deaths and missing persons attributed to flash floods 2000-2022 and their respective shares in the total deaths and missing persons attributed to all flood disasters

3.2 灾害特点

2022 年，灾害总体有 3 个特点。

暴发时间集中。7 月中旬、8 月中旬，黑龙江省五大连池市，四川平武县、北川县、彭州市，山西中阳县，青海大通县相继发生 6 起严重山洪（山洪泥石流）灾害，占全年有人员死亡失踪山洪泥石流灾害事件的 35.3%。

单次灾害伤亡重。四川平武县“7・12”、北川县“7・16”两次山洪泥石流灾害分别造成 18 人死亡失踪，青海大通县“8・18”山洪灾害造成 31 人死亡失踪，3 起山洪（山洪泥石流）灾害共造成 67 人死亡失踪，占全国因山洪（山洪泥石流）灾害死亡失踪人口的 56%。

桥涵阻塞加大灾害损失。黑龙江五大连池市，四川平武县、北川县等山洪（山洪泥石流）灾害，均因沟道桥涵被泥沙、树枝、垃圾等阻塞壅水，导致山洪（山洪泥石流）改道，冲毁房屋、游客露营地，加大灾害损失。

3.3 典型事件

2022 年，全国先后发生了 6 起影响较大的典型灾害事件。

3.3.1 黑龙江省五大连池市“7・12”山洪灾害

7 月 12 日，黑龙江省五大连池市朝阳山镇东风沟（流域面积 35.5 平方千米）突降暴雨，暴雨中心位于距朝阳山镇东风村 9 千米的老公社，推算 4 小时降雨量约 280 毫米、东风沟小流域最大 1 小时面雨量 97 毫米，导致东风沟山洪暴发，洪峰流量约 600 立方米每秒，为东风村有记录以来最大洪水，造成东风村 4 人死亡、1 人失踪。

黑龙江省五大连池市朝阳山镇东风村山洪灾害发生后情景（7 月 16 日）
Dongfeng Village (Chaoyangshan Town of Wudalianchi City, Heilongjiang Province) on July 16, after the flash flood

3.2 Features of the Disasters

Flash flood disasters (flash flood and debris flow) happened in 2022 generally took on the following three characteristics:

The disasters happened in a rather concentrated period of time.In mid July and mid August, six severe flash floods (flash flood and debris flow) hit Wudalianchi City of Heilongjiang Province, Pingwu County, Beichuan County, and Pengzhou City of Sichuan Province, Zhongyang County of Shanxi Province, and Datong County of Qinghai Province, respectively, accounting for 35.3% of all the deadly flash flood and debris flow incidents in 2022.

One single incident could inflict heavy casualties and damage.The "July 12" and "July 16" incidents of Pingwu County and Beichuan County each caused a toll of 18 deaths and missing persons; during the "August 18" flash flood in Datong County, 31 people died or went missing; the life losses of the three incidents totaled 67, accounting for 56% of the national total attributed to flash floods (flash flood and debris flow).

The clogging of bridge openings and culverts had increased the losses. In the cases of Wudalianchi City, Pingwu County and Beichuan County, due to the clogging of watercourses, bridge openings and culverts by silt, tree branches and refuse, the flash floods (flash flood and debris flow) changed courses and swept away houses and camping sites, leading to even heavier losses.

3.3 Major Events

In 2022, six major flash flood (flash flood and debris flow) incidents happened in China.

3.3.1 "July 12" flash flood disaster in Wudalianchi City, Heilongjiang Province

On July 12, 2022, a rainstorm struck the Dongfeng Gully (with a catchment area of 35.5 km^2) in Chaoyangshan Town of Wudalianchi City, Heilongjiang Province. Centered at Laogongshe (9 km from Dongfeng Village of Chaoyangshan Town), the rainstorm brought an estimated 4-hour precipitation of 280 mm and a maximum 1-hour area precipitation of 97 mm at the Dongfeng Gully basin. The heavy downpour triggered flash floods at Dongfeng Gully. With a peak flow of around 600 m^3/s, this worst recorded flooding of Dongfeng Gully killed four and left one missing.

3.3.2 四川平武县“7·12”山洪泥石流灾害

7月11日23时至12日5时，四川平武县木座藏族乡黑水沟（流域面积103平方千米）上游发生短历时强降雨，推算流域6小时面雨量约100毫米，洪峰流量约510立方米每秒。12日5时左右，黑水沟发生山洪泥石流，大量泥石冲入下游河道，导致沟道泥石淤积和桥涵阻塞，河道水位迅速抬高，山洪泥石流改道冲入木座藏族乡场镇和村组，造成7人死亡、11人失踪。

3.3.3 四川北川县“7·16”山洪泥石流灾害

7月15日23时起，四川北川县白什乡及其上游青片河流域持续降雨，白什站最大6小时降雨量69毫米。受降雨影响，16日白什乡上游场镇双溪站洪峰流量279立方米每秒。16日5时20分左右，青片河支流白水河支沟七星沟（流域面积70平方千米）发生泥石流，距离沟口约515米处的拦渣坝（设计库容9400立方米，估算壅水后最大库容约50000立方米）被泥石流淤堵后溃决，拦蓄的大量块石、泥沙、漂木顺沟而下，冲入青片河河道，叠加德州桥阻塞抬高水位影响，大量泥水砂石漫溢涌入白什乡场镇两岸街道和房屋，造成8人死亡、10人失踪，其中6名外地游客在河心德州广场露营时被泥石流冲走。

四川北川县白什乡场镇拦渣坝溃决后情景（7月16日）
The collapsed tailings dam of Baishi Town, Beichuan County, Sichuan Province (July 16)

3.3.2 "July 12" flash flood and debris flow disaster in Pingwu County, Sichuan Province

From 23:00 of July 11 to 5:00 of July 12, Heishui Gully (with a catchment area of 103 km^2, located at Muzuo Xizang Autonomous Town of Pingwu County, Sichuan Province) experienced intensive downpours during a short period of time. Within the Heishui Gully basin, the 6-hour area precipitation was estimated at around 100 mm, and the peak flow was about 510 m^3/s. At 5:00 of July 12, the gully was hit by debris flow. Huge amount of mud and rocks were washed down into the downstream watercourses, clogging the watercourses, bridge openings and culverts, and rapidly lifting the water level. As a result, the flash flood and debris flow changed course and gushed into the residential areas of Muzuo Xizang Autonomous Town, killing 7 and leaving 11 missing.

3.3.3 "July 16" flash flood and debris flow disaster in Beichuan County, Sichuan Province

Starting from 23:00 of July 15, persistent rainfall was experienced in Baishi Town of Beichuan County, Sichuan Province, and the upstream Qingpian River basin. The maximum 6-hour precipitation was 69 mm at Baishi Station. As a result, on July 16, the peak flood flow reached 279 m^3/s at Shuangxi Station (located at the upstream settlement of Baishi Town). At around 5:20 of July 16, Qixing Gully (with a catchment area of 70 km^2), a branch of Baishui River (a tributary of Qingpian River), was hit by debris flow, followed by clogging and then the collapse of a tailings dam (with a designed storage capacity of 9,400 m^3, and a estimated maximum capacity of 50,000 m^3) 515 m away from the gully mouth. As a result, huge amount of rocks, silt and driftwood were washed down into Qingpian River. Worse still, the water under the Dezhou Bridge was clogged, further lifting the water level. Eventually, the debris overflew into the streets and dwellings of Baishi Town, killing 8 and leaving 10 missing, of which 6 were tourists washed away by the debris flow when camping at the Dezhou Square in the middle of the river channel.

3.3.4 山西中阳县“8·11”山洪灾害

8月11日12时许，山西中阳县鼻子亩沟（流域面积2平方千米）发生山洪，山洪冲入钢筋加工场工棚，造成离隰高速第一标段项目部5人死亡。据调查，距事发点1千米的柳林县贺家岭村12小时（10日23时至11日11时）降雨量205毫米，鼻子亩沟钢筋加工场断面洪峰流量为41立方米每秒。

山西省中阳县鼻子亩沟钢筋加工场山洪灾害发生后情景（8月11日）
The steel bar processing workshop (Bizimu Gully, Zhongyang County, Shanxi Province) on August 11, after the flash flood disaster

3.3.5 四川彭州市龙槽沟“8·13”山洪灾害

龙槽沟位于四川彭州市龙门山镇小鱼洞社区和宝山村交界处，流域面积16.1平方千米，近年来成为网红打卡地。8月13日14时，龙槽沟上游发生历时约1小时降雨，推算流域面雨量约17毫米、洪峰流量18立方米每秒。因暑期天气持续高温，大量游客进入龙槽沟戏水纳凉，15时30分水位突然上涨，部分人员未及时撤离，造成7人死亡。

3.3.4 "August 11" flash flood disaster in Zhongyang County, Shanxi Province

At 12:00 of August 11, a flash flood struck Bizimu Gully (with a catchment area of 2 km^2) of Zhongyang County, Shanxi Province. The flood gushed into the work shed of a steel bar processing workshop, killing 5 workers of the Lishi-Xixian Expressway Project Section I. Investigation indicates the 12-hour precipitation (from 23:00 of August 10 to 11:00 of August 11) at Hejialing Village of Liulin County (1 km from the incident site) reached 205 mm, and the section peak flood flow at the workshop was 41 m^3/s.

3.3.5 "August 13" flash flood disaster at Longcao Gully, Pengzhou City, Sichuan Province

Longcao Gully sits at the border between Xiaoyudong Community and Baoshan Village of Longmenshan Township, Pengzhou City, Sichuan Province, and has a catchment area of 16.1 km^2. In recent years, the Gully has become a popular summer camping site fanned by internet influencers. At 14:00 on August 13, rainfall hit its upper reaches and lasted for around an hour. The area precipitation over the basin was estimated at 17 mm, and the peak flood flow at 18 m^3/s. Due to sustained high temperatures, many people went to the gully to avoid summer heat. The abrupt rise of the water at 15:30 left little withdrawal time for the tourists. Seven died as a result.

四川彭州市龙槽沟山洪灾害发生时情景（8 月 13 日）
Longcao Gully (Pengzhou City, Sichuan Province) on August 13, when the flash flood hit

3.3.6 青海大通县“8·18”山洪灾害

8月17日22—23时，青海大通县沙岱河（流域面积124平方千米）和临近的下湾沟（流域面积3.2平方千米）发生强降雨，暴雨中心青山乡沙岱村五组站降雨量120毫米，引发山洪，洪峰流量188立方米每秒，洪水重现期超500年。山洪漫溢冲毁青山、青林2乡贺家庄、沙岱、生地、棉格勒等村房屋及临时放牧点，造成26人死亡、5人失踪。

3.4 防御工作

2022年，水利部及地方各级水利部门积极主动作为，超前安排部署，强化预报、预警、预演、预案“四预”措施，全力防范化解山洪灾害风险，最大程度减少了山洪灾害人员伤亡和灾害损失。

青海大通县沙岱河两岸山洪灾害发生后情景（8月18日）
Flood-stricken villages along the Shadai River, Datong County, Qinghai Province (August 18)

3.3.6 "August 18" flash flood disaster in Datong County, Qinghai Province

From 22:00 to 23:00, August 17, intensive rainfall hit Shadai River (with a catchment area of 124 km^2) and the nearby Xiawan Gully (with a catchment area of 3.2 km^2) of Datong County, Qinghai Province. At the Wuzu Station of Shadai Village, Qingshan Town, which was the center of the rainstorm, the precipitation was recorded at 120 mm. The heavy rainfall triggered a flash flood, with a peak flood flow of 188 m^3/s and a 500-year return period. The floodwater devastated dwellings and temporary grazing spots at the Hejiazhuang, Shadai, Shengdi, and Miangele villages of Qingshan Town and Qinglin Town, leaving 26 dead and 5 missing.

3.4 Prevention and Control

In 2022, MWR and local water resources departments made proactive and forwardlooking arrangements, and enhanced the four preemptive pillars of forecasting, early warning, exercising, and contingency planning. Thanks to the all-out effort in preventing and controlling flash flood hazards, the casualties and losses attributed to flash flood disasters had been minimized.

3.4.1 工作部署

3月，水利部制定印发《关于加强山洪灾害防御工作的指导意见》，以场次降雨过程为主线制定了省、市、县3级山洪灾害监测预警工作清单，进一步规范部署山洪灾害防御工作。4月26日，召开全国山洪灾害防御工作视频会议，专题安排部署全年山洪灾害防御重点工作。汛期，先后3次印发通知，专题部署山洪灾害监测预警、提前转移避险等工作。

3.4.2 风险隐患排查整治

3—5月，组织全国有山洪灾害防治任务的29省（自治区、直辖市）和新疆生产建设兵团，统筹新型冠状肺炎疫情防控，以山洪灾害自动监测系统运行管理、监测预警平台运行、预案修订更新、责任制落实、山洪沟道侵占等为重点，全面深入排查山洪风险隐患，建立问题清单、整改清单、责任清单，确保汛前整改到位。各地共排查整治山洪风险隐患27008处，其中监测设施设备方面7154处、监测预警平台运行方面2562处、群测群防体系方面15389处、山洪沟道隐患方面1903处。

3.4.3 监督检查

5—8月，组织开展山洪灾害监测预警监督检查，共检查县级监测预警平台137个（新检查105个、“回头看”复查32个）、自动监测站点480个，发现问题124个，针对发现的问题督促各地按期完成整改。7—9月，组织各地对照省级监测预报预警平台技术要求，就平台建设、功能实现情况开展自查自改自评估，针对自查发现的问题，建立问题整改清单，制定整改提升措施。10—12月，在各地自查自改自评估基础上，水利部逐一对省级山洪灾害监测预报预警平台功能进行线上检查，针对发现的问题，以“一省一单”形式督促限期改进提升。

3.4.1 Arrangements

In March, 2022, MWR drafted and circulated *Guiding Opinions on Enhancing Flash Flood Prevention and Control*, devising the precipitation-process-based lists of flash flood monitoring and early warning tasks at the provincial, municipal, and county levels, and further standardizing and charting out the prevention and control tasks. On April 26, MWR held a national video conference to map out the priority tasks of flash flood prevention and control for 2022. During the flood season, three Circulars were issued for the implementation of flash flood monitoring and early warning, and early evacuation and relocation.

3.4.2 Risk factors investigation and correction

From March to May, MWR launched a campaign of flash flood hazards investigation and derisking in 29 provinces/autonomous regions/municipalities and Xinjiang Production and Construction Corps, which are tasked with flash flood prevention and control. Coordinated with the prevention and control of the COVID-19 pandemic, the campaign focused on the operation and management of the automatic flash flood monitoring system and the early warning platform, the revision and update of contingency plans, the identification of responsible officials, and the encroachment of flash flood paths. Following the checklists of problems and the accountability system established, due rectifications were made before the flood season. In total, 27,008 flash flood risk factors were identified and eliminated, of which 7,154 concerned monitoring facilities and equipment, 2,562 were about the operation of the monitoring and early warning platform, 15,389 were related to the grassroots-based monitoring and prevention system, and 1,903 concerned the hazard-prone gullies.

3.4.3 Supervision and inspection

From May to August, MWR organized the supervision and inspection of flash flood monitoring and early warning, covering 137 county-level monitoring and early warning platforms (105 are first-time checks and 32 are rechecks), and 480 automatic monitoring stations. All together 124 issues were identified, and MWR urged and guided immediate and due rectification. From July to September, a round of self-inspection, self-rectification and self-evaluation was carried out to track the construction and functioning of monitoring, forecasting and early warning platforms against the provincial-level technical standards; checklists of problems were drawn, and rectification measures were developed. From October to December, based on the previous self-inspection, self rectification and self evaluation, MWR conducted online testing of the functions of each and every provincial flash flood monitoring, forecasting and early warning platform, issued problem checklists, and urged rectification and improvement within a prescribed time frame.

3.4.4 预警发布

水利部会同中国气象局制作发布山洪灾害气象预警 140 期，其中中央电视台播出 43 期，并“点对点”发至地方。建立山洪灾害监测预警抽查日报机制，汛期累计抽查强降雨覆盖省份县级山洪灾害监测预警情况 2550 县次。全国各地共利用山洪灾害监测预警系统发布县级山洪灾害预警信息 18 万次，向 786 万人次（防汛责任人）发送预警短信 4422 万条，启动无线预警广播 38.6 万次，依托“三大运营商”向社会公众发布预警短信 19.4 亿条，为受山洪威胁地区群众及时转移避险提供了有力支撑。

3.4.5 项目建设管理

2022 年，中央财政共安排水利发展资金 23 亿元，支持地方开展山洪灾害补充调查评价、省级监测预警平台巩固提升、简易监测预警设施设备配备等山洪灾害防治非工程措施建设及运行维护，实施 183 条重点山洪沟防洪治理，其中专门安排 3 亿元支持浙江等 9 省（自治区）在重点地区开展小流域山洪灾害防御能力提升项目建设。水利部始终坚持建管并重，及时下达山洪灾害防治年度项目建设任务和中央补助资金，印发项目建设工作要求，制定技术文件，定期统计通报项目建设进展。组织召开 2 次视频会议，安排部署山洪灾害防御能力提升项目建设，分析项目建设重点、难点、堵点，加快推进项目建设，确保按时保质完成目标任务。

3.5 防御成效

2022 年，各地充分利用已建山洪灾害监测预警系统和群测群防体系，强化预警信息发布，及时转移受威胁群众，努力实现监测精准、预警及时、反应迅速、转移快捷、避险有效，最大程度保障了人民群众生命安全。2022 年因山洪灾害死亡失踪人口为有统计资料以来最低，较 2011—2021 年平均死亡失踪人口下降 65.5%，山洪灾害防御成效显著。

3.4.4 Issuing early-warning information

In collaboration with China Meteorological Administration (CMA), MWR issued 140 flash flood meteorological announcements (43 of which were aired on China Central Television) and promptly notified relevant localities. By setting up the random check and daily report mechanism, it conducted 2,550 random checks of the implementation of county-level flash flood monitoring and early warning in provinces experiencing heavy precipitation during the flood season. Through the flash flood monitoring and early warning system, a total of 180,000 flash flood alert messages were issued at county-level, 44.22 million alerts were sent to 7.86 million officials responsible for flood control, 386,000 early warning wireless broadcasts were made, and 1.94 billion alert messages were sent to the general public with the assistance of China Mobile, China Telecom and China Unicom, which had underpinned the evacuation and relocation of people threatened by flash floods.

3.4.5 Project construction and management

In 2022, the Central Government allocated 2.3 billion RMB of water conservancy development fund for the construction, operation and management of non-engineering flash flood prevention and control measures and facilities by the local governments (including supplemental flash flood investigation and assessment, the consolidation and upgrading of provincial monitoring and early warning platforms, and the equipment of simple monitoring and early warning tools), and the treatment of 183 major hazard-prone gullies. Of the total sum, 300 million RMB were directed towards the key areas of Zhejiang and 8 other provinces/autonomous regions for projects of improving flash flood prevention and control capacity in small basins. Laying equal stress on construction and management, MWR issued in a timely manner the annual construction tasks of flash flood prevention and control project and appropriated the relevant subsidies from the Central Government without delay, circulated the requirements concerning project construction, formulated technical papers, and regularly collected information and reported on project progress. It also held two video conferences to map out the flash flood prevention and control capacity improvement projects and analyze their priorities, difficulties and bottlenecks, so as to accelerate project construction and ensure the completion of all tasks in a timely and quality-assured manner

3.5 Effectiveness of Flash Flood Disaster Prevention

In 2022, local governments made full use of the flash flood monitoring and warning systems and mass monitoring and prevention systems. Early warning release and prompt evacuation were strengthened. Maximum efforts were made to protect lives through accurate monitoring, timely warning, rapid response, quick evacuation, and effective risk avoidance. The number of deaths and missing persons attributed to flash flood disasters in 2022 was down by 65.5% compared with 2011–2021 annual average and reached a historic low.

案例 1 浙江龙泉市城北乡东书村“6·20”山洪灾害预警避险

6 月 20 日，浙江龙泉市城北乡东书村西溪自然村突发降雨，最大 6 小时累计降雨量 124 毫米，暴雨重现期约 20 年，导致山洪暴发，水位暴涨，洪水最大时淹没路面 1～2 米，部分房屋严重进水。

浙江省充分发挥“预报预警、监测预警、现地预警”互为补充的渐进式预警体系作用，依托省级山洪联防平台，科学预见了当地山洪风险，及时发出预警。城北乡在龙泉市水利局预警叫应后，迅速启动防汛应急Ⅰ级响应。东书村村支部委员会和村居民委员会第一时间赶到西溪自然村巡查排险，发现河道水位在 5 分钟内迅猛上涨，立即安排村干部进村入户组织全村 20 户 46 人全部转移。半小时后，山洪侵袭村庄。由于预警及时、叫应有效、转移到位，全村无一人伤亡。

浙江龙泉市东书村西溪自然村受山洪冲击时情景（6 月 20 日）
Flood struck the Xixi Community of Dongshu Village, Longquan City, Zhejiang Province (June 20)

Case 1 Early warning and evacuation during the “June 20” flash flood in Dongshu Village, Chengbei Town, Longquan City, Zhejiang Province

On June 20, intensive rainfall hit the Xixi community of Dongshu Village, Chengbei Town (Longquan City, Zhejiang Province). With a maximum 6-hour precipitation of 124 mm and a 20-year return period, the heavy downpour triggered flash flood and water level surge. At its peak, the flood inundated the streets by 1-2 m above the ground. Some dwellings were severely flooded.

Tapping into the step-by-step early warning system incorporating forecast, monitoring and on-site warning, and the provincial joint platform of flash flood prevention, the Water Resources Bureau of Longquan City foresaw the flash flood hazard and issued early warning. Chengbei Town immediately activated the highest level of flood control response. Receiving the warning, Dongshu Village organized patrols and found the water level of the river had risen rapidly within 5 minutes. In response, all of the 46 people of the 20 households to be affected were evacuated. 30 minutes after the evacuation, flood struck the village. Thanks to the timely warning and responses, casualties were avoided.

案例 2 广西桂林市龙胜县“6・22”山洪灾害紧急避险

6 月 16—22 日，广西桂林市龙胜县遭遇持续强降雨，42 个站点降雨量超过 250 毫米，累计最大降雨量 768 毫米，22 日 16 时 10 分龙胜镇玉龙巷后山暴发山洪泥石流灾害。

6 月 21 日，广西壮族自治区水利厅和气象局联合发布山洪灾害气象预警，预报龙胜镇等 8 个乡（镇）、17 个危险区发生山洪灾害的可能性大，提醒影响区做好监测、巡查、预警和转移避险等防范工作。6 月 22 日，按照山洪灾害防御预案，龙胜镇及时果断采取敲门呼叫、入户寻找等方式，组织周边 4 栋房子共 32 人全部安全转移，并继续组织泥石流上下方其他住户进行“延伸式”安全转移，避免了人员伤亡。

广西桂林市龙胜县山洪灾害发生后情景（6 月 22 日）
The flash flood-stricken Longsheng County (Guilin City, Guangxi Zhuang Autonomous Region) on June 22

Case 2 Emergency evacuation during the “June 22” flash flood disaster in Longsheng County, Guilin City, Guangxi Zhuang Autonomous Region

From June 16 to June 22, Longsheng County (Guilin City, Guangxi Zhuang Autonomous Region) experienced sustained heavy downpour. 42 local stations recorded precipitations over 250 mm, and the maximum cumulative precipitation reached 768 mm. At 16:10, June 22, flash flood and debris flow struck at the mountain behind the Yulong Alley, Longsheng Town.

On June 21, the Water Resources Department and Meteorological Bureau of Guangxi jointly issued warning against the high probability of flash flood in 8 towns (including Longsheng Town) and 17 danger zones, and called for close monitoring, patrolling and early warning, and preparation for evacuation and relocation in hazard-prone regions. On June 22, according to the contingency plan, local officials of Longsheng Town went from door to door to urge for evacuation, and managed to evacuate all the 32 residents safely. People living upstream and downstream from the hazard site were also urged to evacuate in caution. No casualties were caused.

4 干旱灾害防御

DROUGHT DISASTER PREVENTION

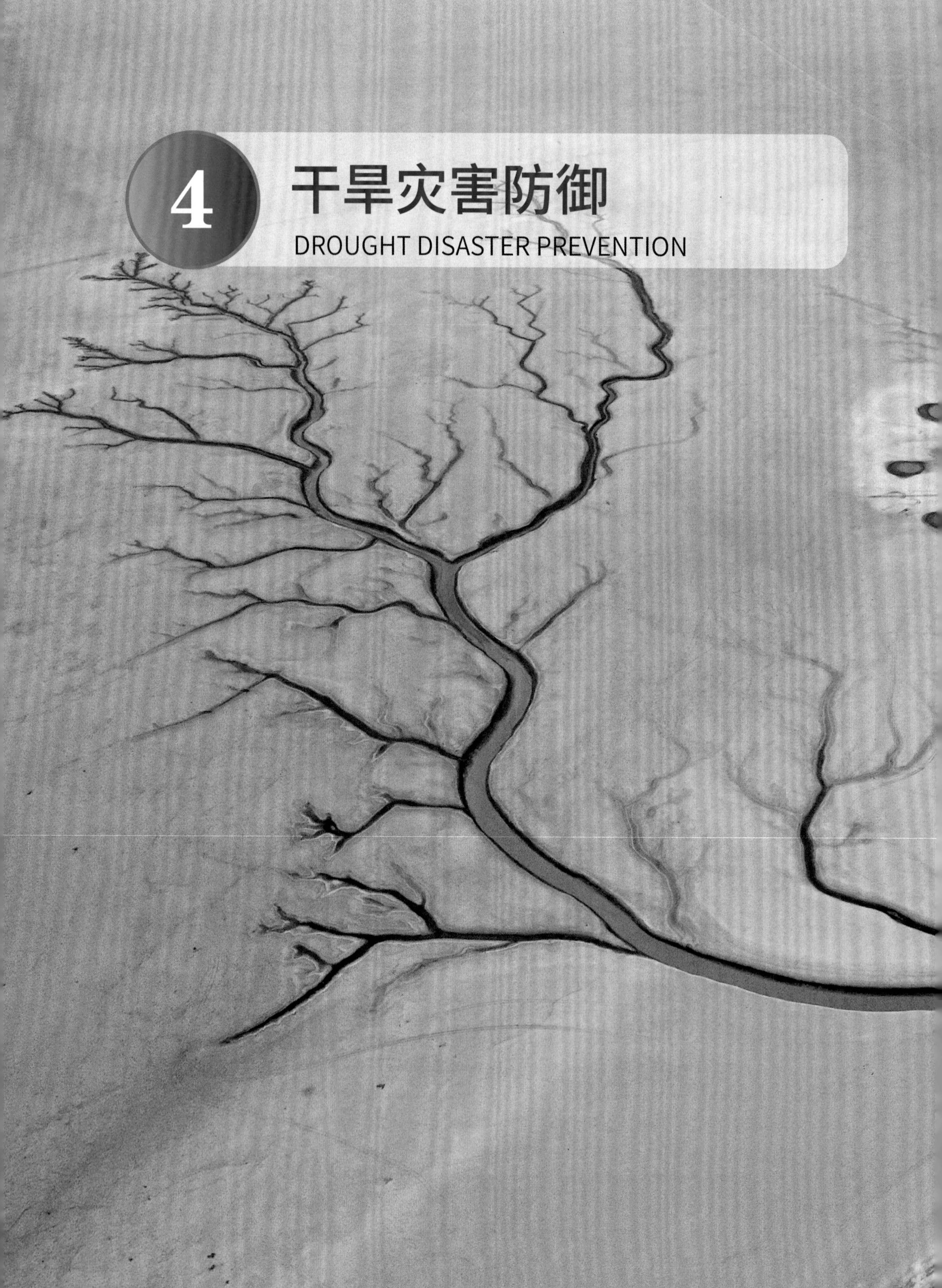

4.1 旱情及特点

2022 年，全国平均降水量较常年略偏少，时空分布严重不均。珠江流域发生冬春连旱，黄淮海和西北地区发生春夏连旱，长江流域发生夏秋连旱，全国有 24 省（自治区、直辖市）发生不同程度干旱，旱情总体有 4 个特点。

4.1 Droughts and the Characteristics

In 2022, the national average precipitation was slightly lower than normal, and its spatial and temporal distribution was notably uneven. The Pearl River basin suffered from prolonged droughts in winter and spring; the North China Plain and the Northwest China went through a dry spell lasting from summer to autumn; 24 provinces/autonomous regions/municipalities nationwide experienced different levels of water shortage. In general, drought in 2022 took on the following four characteristics.

受旱范围广，丰水地区旱情重。2022 年，长江、黄河、淮河、海河、珠江和太湖六大流域均发生干旱，其中长江、珠江、太湖流域等丰水地区气象水文干旱严重。长江流域夏秋期间，平均气温之高、高温日数之多、降雨量之少、江湖水位之低，均为 1961 年有实测记录以来同期之最；珠江流域中东部遭遇 60 年来最严重旱情，旱情自 2021 年冬持续至 2022 年春，粤港澳大湾区、粤东、闽南等地供水安全受到威胁；太湖流域及东南诸河遭遇罕见的夏秋连旱，多站水位一度位列有实测记录以来最低。

长江珠江咸潮发生早，持续时间长。受长江干流持续来水偏少和连续 3 个台风（2211 号“轩兰诺”、2212 号“梅花”、2214 号“南玛都”）助推影响，9 月上旬长江口遭受罕见咸潮入侵，较往年提前 3 ～ 4 个月。上海市水源地陈行、青草沙、东风西沙水库取水口受咸潮影响时间长达 65 天、87 天和 80 天，上海市供水安全受到严重威胁。受上游来水偏枯及 2220 号台风“纳沙”影响，珠江口咸潮影响偏早偏强，珠海市平岗、竹洲头等主力泵站取水口氯化物含量超标，较常年分别偏早 25 天、30 天；中山市全禄、大丰等主力水厂取水口氯化物含量超标，较常年偏早近 2 个月；东江三角洲 2021 年 12 月底至 2022 年 2 月初咸潮影响持续 34 天。

The impact of drought was extensive, and regions traditionally rich in water resources suffered severely.In 2022, drought struck six major basins of the Yangtze River, the Yellow River, the Huaihe River, the Haihe River, the Pearl River, and the Taihu Lake. The traditionally water-abundant Yangtze River basin, the Pearl River basin, and the Taihu Lake basin all experienced severe hydrological and meteorological droughts. For the Yangtze River basin, the summer and autumn of 2022 witnessed the highest average temperature, the most high temperature days, the leanest precipitation, and the lowest water level of rivers and lakes over the same period since records began in 1961. The central-eastern Pearl River basin was stricken by the worst drought in 60 years. The drought lasted from the winter of 2021 to the spring of 2022, threatening water supply in the Greater Bay Area, the eastern Guangdong Province, and the southern Fujian Province. The Taihu Lake basin and smaller rivers in the southeastern China went through prolonged droughts in summer and autumn, something rarely seen in this region, and many stations reported record-low water levels.

Saltwater intrusion set in early and lasted long at the estuaries of the Yangtze River and the Pearl River. Due to sustained less-than-normal water inflow in the mainstream and the impact of three consecutive typhoons–Typhoon “Hinnamnor” (No. 2211), Typhoon “Muifa” (No. 2212), and Typhoon “Nanmadol” (No. 2214), the Yangtze River estuary experienced saltwater intrusion in early September, 3-4 months earlier than in previous years. The water intakes of Chenhang Reservoir, Qingcaosha Reservoir, and Dongfengxisha Reservoir (supplying water to Shanghai) were affected by the saltwater intrusion for 65 days, 87 days, and 80 days, respectively, gravely threatening the city’s water supply. Also due to the less-than-normal river water inflow and the impact of Typhoon “Nesat” (No. 2220), the Pearl River estuary faced a similar situation. At the water intakes of Pinggang and Zhuzhoutou Pumping Stations in Zhuhai City, excessive chloride content was detected at dates 25 and 30 days earlier than in previous years. At Quanlu, Dafeng and other major water plants of Zhongshan City, the same happened nearly 2 months earlier than in previous years. From late December, 2021 to early February, 2022, the Dongjiang River delta was affected by saltwater intrusion for 34 consecutive days.

局地水源不足，城乡供水和农业灌溉受威胁。广州、东莞等市局部地区受珠江口咸潮影响出现短时降压供水、供水口感略咸等情况。受长江口咸潮和长江低水位影响，上海、南昌、武汉等沿江城市采取应急供水措施保障城市供水安全。长江流域部分以小型水库或山泉水、溪流水为水源的山丘区，一度有 81 万人、92 万头大牲畜因旱临时饮水困难。长江流域旱情发生在中稻孕穗至抽穗扬花期、晚稻返青至分蘖期和夏玉米抽雄至灌浆期，黄淮海和西北地区春夏旱发生在北方冬小麦拔节、抽穗、灌浆期和夏玉米、夏大豆播种期，用水需求大，对粮食生产造成严重威胁。

干旱影响面广，波及生态发电航运。长江流域受旱期间，长江干流河道水位持续走低，7—10 月出现历史同期最低水位，荆江太平口、藕池口提前近 3 个月断流，洞庭湖和鄱阳湖提前 3 个多月进入枯水期，鄱阳湖湖面面积一度仅为历史同期均值的 1/10，洞庭湖水体面积 10 月 1 日减少至 309.9 平方千米（1998 年以来同期最小值），湖泊和部分河道水质急剧恶化，生态系统严重受损。旱情严重时，四川省水电日发电能力从 9 亿千瓦时减少至 4.4 亿千瓦时，下降了 51.1%，部分地区生产生活用电受到严重影响。因长江枯水期提前，长江航道黄金航运期提前结束，航道等级下降，长江航运承载能力一度减少近 20%。

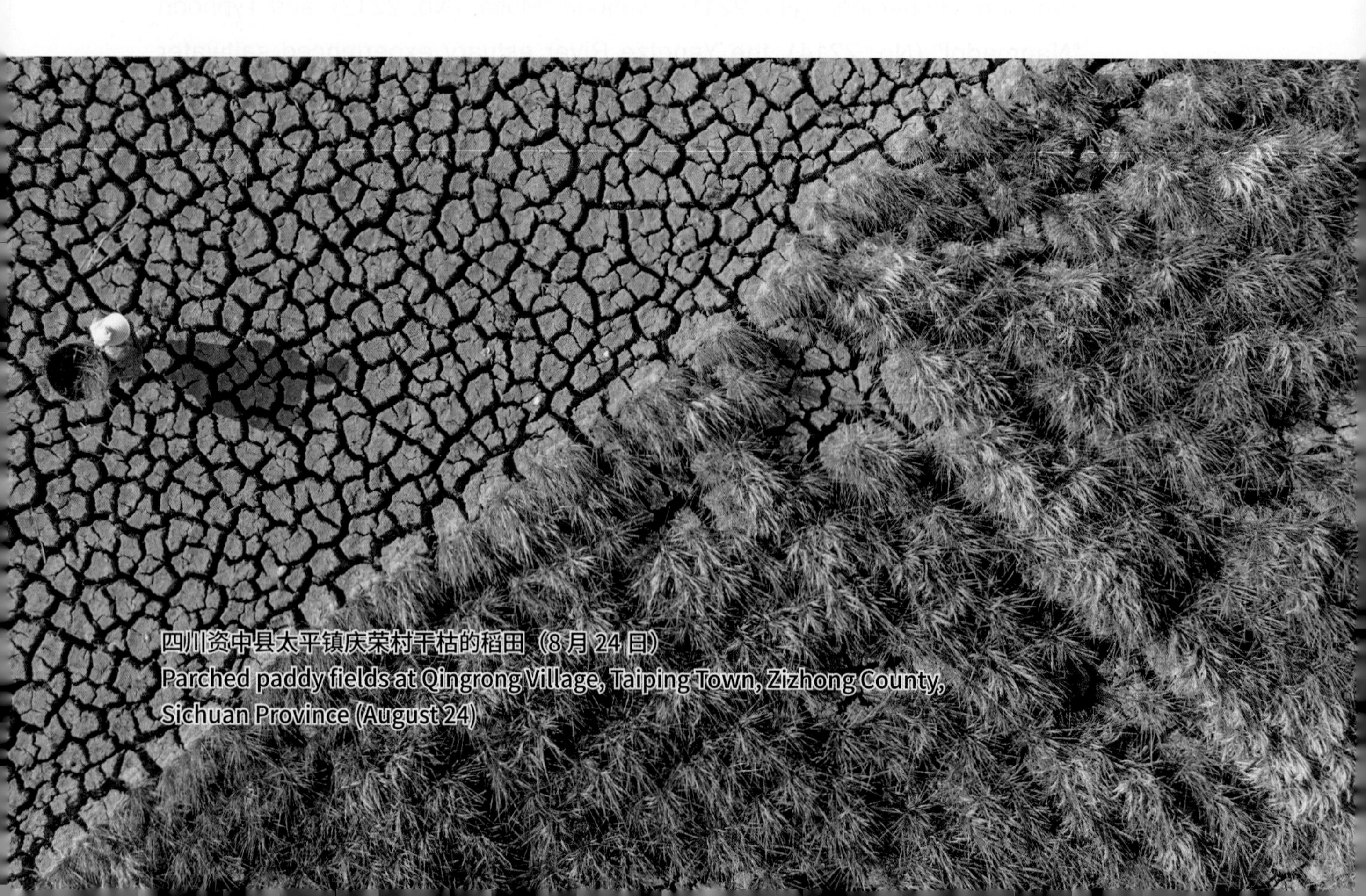

四川资中县太平镇庆荣村干枯的稻田（8 月 24 日）
Parched paddy fields at Qingrong Village, Taiping Town, Zizhong County, Sichuan Province (August 24)

Localized water shortage threatened urban water supply and agricultural irrigation.Due to the saltwater intrusion at the Pearl Rivers estuary, cities such as Guangzhou and Dongguan temporarily experienced lowered water supply pressure and slightly salty drinking water. Facing the impact of low Yangtze River water level and saltwater intrusion, cities such as Shanghai, Nanchang and Wuhan adopted emergency measures to guarantee urban water supply. In some hilly areas within the Yangtze River basin that rely on small reservoirs and mountain springs and streams for water supply, there were occasions when 810,000 people and 920,000 livestock experienced difficulties accessing drinking water. When drought hit the Yangtze River basin, it was the panicle booting and flowering stage for semi-late rice, the regreening and tillering stage for late-season rice, and the tassling and grain-filling stage for maze; when the North China plain and the northwest China went through the dry spring and summer, it was the jointing, tillering, and grain-filling stage for winter wheat, and the seeding season for summer maze and soybean. During this critical period, crops need large amount of water. Therefore, the drought posed grave threats to food production.

The impact of the drought had spread to other sectors such as eco-system, power generation and navigation. During the drought, the water level stayed low in the Yangtze River mainstream, and the four months from July to October witnessed the lowest water level over the same period in history. At Taipingkou and Ouchikou, the Jingjiang River (a tributary of the Yangtze River) dried up nearly 3 months earlier than in previous years. Dongting Lake and Poyang Lake entered dry season 3 months earlier than normal, the surface area of Poyang Lake was at one point only 1/10 of its historical average, and Dongting Lake shrank to 309.9 km^2 (the smallest over the same period since 1998) on October 1. With drastically deteriorating water quality in lakes and some river channels, the eco-system suffered severely. In the gravest circumstances, the daily hydropower generation capacity of Sichuan province dropped from 900 million kWh to 440 million kWh. This 51.1% decline led to curtailed household and industry power supply in some regions. As the dry season set in earlier, the golden time of Yangtze River navigation ended prematurely. As the waterway was downgraded, the navigation capacity of the river was at one point reduced by 20%.

4.2 主要干旱过程

4.2.1 珠江流域冬春连旱

2022 年 1 月，珠江流域东江、韩江流域降水量较常年同期偏少 3 ～ 5 成，骨干水库有效蓄水率仅为 8% 和 19%，新丰江水库在死水位以下运行 25 天，同时受珠江口咸潮偏强影响，呈“冬春连旱、旱上加咸”态势，珠江三角洲、汕头、梅州等地延续上年旱情并持续发展，有 13 万农村人口发生饮水困难，181 万城镇人口和 97 万农村人口供水受影响。2—3 月，旱区出现多次有效降雨，旱情逐步解除。

4.2.2 黄淮海和西北地区春夏连旱

4 月至 6 月中旬，黄淮海和西北大部降水量较常年同期偏少 3 ～ 7 成，气温较常年同期偏高 1 ～ 3°C，部分地区土壤中度以上缺墒，旱情快速发展。旱情高峰时，作物受旱面积 4358.67 千公顷，有 16 万人、146 万头大牲畜因旱发生饮水困难，主要分布在河北、山西、内蒙古、安徽、山东、河南、陕西、甘肃等省（自治区），其中河南、内蒙古两省（自治区）旱情最重。6 月下旬至 7 月上旬，旱区出现较强降雨过程，大部分地区旱情陆续解除，内蒙古自治区西部旱情持续至 10 月。

4.2.3 长江流域夏秋连旱

7—10 月，长江流域发生了 1961 年有实测记录以来最严重的气象水文干旱。一是降雨少。流域累计降雨量 291 毫米，较常年同期偏少 39%，为 1961 年以来同期最少。二是气温高。高温日数长达 45.6 天，亦为 1961 年以来同期之最。三是来水枯。流域来水较常年同期总体偏少超过 4 成，其中洞庭湖水系偏少 6 ～ 7 成、鄱阳湖水系偏少 7 ～ 9 成，8 月中下游干流出现超百年一遇枯水。四是水位低。中下游干流及洞庭湖、鄱阳湖水位均创有实测记录以来同期最低水位，鄱阳湖星子站 9 月 23 日跌破历史最低水位。旱情高峰时，作物受旱面积 4421.33 千公顷，有 81 万人、92 万头大牲畜因旱发生饮水困难，主要分布在四川、重庆、湖北、湖南、安徽、江西、河南、贵州、江苏、陕西等省（直辖市）。8 月下旬至 10 月，长江上游、汉江上游等地发生强降雨过程。10 月，旱区大部旱情逐步缓解，但重庆、湖南两省（直辖市）局地群众饮水困难情况持续至 12 月。

4.2 Major Drought Processes

4.2.1 Winter-spring drought in the Pearl River basin

In January, 2022, precipitation in the Dongjiang River basin and Hanjiang River basin (sub-basins of the Pearl River) was 30%-50% lower than normal. In the two basins, the effective storage ratios of critical reservoirs were only 8% and 19%, respectively, and in the case of Xinfengjiang Reservoir, it went through 25 days operating below the dead storage level. Meanwhile, as the winter drought was extended into spring, the saltwater intrusion at the Pearl River estuary also weighed in. The double impact of prolonged water shortage and increasing water salinity was felt in the Pearl River delta, Shantou City, and Meizhou City, where 130,000 rural residents experienced difficulties accessing drinking water, and 1.81 million urban residents and 970,000 rural residents had their water supplies affected. During February and March, the drought was eventually subdued by several rounds of effective rainfall.

4.2.2 Spring-summer drought of the North China Plain and Northwestern China

From April to mid June, most of the North China Plain and Northwestern China was 30%-70% drier and 1-3 °C warmer than normal, leading to low or severely low soil moisture. Drought unfolded rapidly, and at its peak 4,358,670 ha of cropland were affected, and 160,000 people and 1.46 million bigger-sized livestock experienced difficulties accessing drinking water. Most of this happened in the provinces/autonomous regions of Hebei, Shanxi, Inner Mongolia, Anhui, Shandong, Henan, Shaanxi, and Gansu, and among them Henan and Inner Mongolia were the worst hit. In late June and early July, the drought was subdued in most of the affected regions after moderate to heavy rainfall. The dry spell in the western part of Inner Mongolia lasted through October.

4.2.3 Summer-autumn drought in the Yangtze River basin

From July to October, the Yangtze River basin witnessed the severest meteorological and hydrological drought since measured records began in 1961. First, less rainfall was received. The cumulative precipitation was only 291 mm, the lowest over the same period since 1961, and 39% less than normal. Second, it was hotter. The number of high temperature days reached 45.6, the most over the same period since 1961. Third, water inflow was insufficient. In general, water inflow was 40% below the normal level; in the Dongting Lake basin, it was 60%-70% lower; in the Poyang Lake basin, the shortfall was as large as 70%-90%; in mid and late August, the middle and lower reaches of the Yangtze River mainstream suffered from extremely small inflow (with an over-100-year return

鄱阳湖中的千年石岛落星墩完全露出（8 月 21 日）
Luoxingdun, a thousand-year-old stone islet in Poyang Lake, was completely exposed (August 21)

4.3 干旱灾情

2022 年，全国 24 省（自治区、直辖市）发生干旱灾害。农作物因旱受灾面积 6090.21 千公顷，比前 10 年的平均值下降 29.6%；绝收面积 611.78 千公顷，比前 10 年的平均值下降 32.3%；因旱粮食损失 57.44 亿千克，比前 10 年的平均值下降 59.4%；经济作物损失 149.44 亿元；542.38 万人因旱饮水困难，比前 10 年的平均值下降 43.8%；331.92 万头大牲畜因旱发生饮水困难，比前 10 年的平均值下降 48.2%。

period). Fourth, water levels were low. The middle and lower reaches of the mainstream, Dongting Lake, and Poyang Lake all witnessed record low levels. On September 23, the Xingzi Station recorded the lowest water level in history. When the drought reached its peak, 4,421,330 ha of cropland were affected, and 810,000 people and 920,000 bigger-sized livestock experienced difficulties accessing drinking water. Most of this happened in the provinces/municipality of Sichuan, Chongqing, Hubei, Hunan, Anhui, Jiangxi, Henan, Guizhou, Jiangsu, and Shaanxi. From late August to October, the upper Yangtze and the Hanjiang River received heavy rainfall. By October, the drought had been mostly subdued. However, in parts of Chongqing and Hunan, the drinking water disruptions lasted through December.

4.3 Disasters and Losses

In 2022, 24 provinces/autonomous regions/municipalities nationwide suffered drought disasters. A total of 6,090,210 ha of cropland were affected, 29.6% less than the preceding decadal average; of all the affected cropland, 611,780 ha suffered crop failure, 32.3% lower than the preceding decadal average; the grain yield loss attributed to drought was 5.744 billion kg, down by 59.4% from the preceding decadal average; the loss of cash crops amounted to 14.944 billion RMB; a total of 5.4238 million people experienced difficulties accessing drinking water due to drought, 43.8% less than the preceding decadal average; and 3.3192 million bigger-sized livestock experienced difficulties accessing drinking water due to drought, down by 48.2% from the preceding decadal average.

表 4-1　2022 年全国农作物因旱受灾面积、绝收面积情况（单位：千公顷）
Table 4-1　Cropland area affected and failed by drought in 2022 (in 1,000 ha)

地区 Province	农作物因旱受灾面积 Cropland area affected by drought	农作物因旱绝收面积 Cropland area failed by drought	地区 Province	农作物因旱受灾面积 Cropland area affected by drought	农作物因旱绝收面积 Cropland area failed by drought
全国 Nationwide	6090.21	611.78	河南 Henan	602.75	39.19
北京 Beijing			湖北 Hubei	964.60	94.49
天津 Tianjin			湖南 Hunan	691.68	77.29
河北 Hebei	28.14	5.50	广东 Guangdong		
山西 Shanxi	113.42	3.55	广西 Guangxi	113.44	7.91
内蒙古 Inner Mongolia	543.42	50.67	海南 Hainan		
辽宁 Liaoning			重庆 Chongqing	332.22	66.16
吉林 Jilin			四川 Sichuan	522.48	53.67
黑龙江 Heilongjiang			贵州 Guizhou	265.72	26.75
上海 Shanghai			云南 Yunnan	169.26	26.04
江苏 Jiangsu	71.76	4.61	西藏 Xizang	2.95	0.17
浙江 Zhejiang	54.05	6.00	陕西 Shaanxi	280.95	44.06
安徽 Anhui	364.81	14.49	甘肃 Gansu	169.64	5.92
福建 Fujian	26.06	2.28	青海 Qinghai	5.28	0.07
江西 Jiangxi	703.24	80.07	宁夏 Ningxia	26.08	1.13
山东 Shandong	1.64		新疆 Xinjiang	36.62	1.76

注 数据来源于应急管理部，空白表示无灾情。

Note The data come from the Ministry of Emergency Management; spaces in blank denote no such losses by drought.

表 4-2 2022 年全国因旱饮水困难情况
Table 4-2 Difficulties accessing drinking water attributed to drought nationwide in 2022

地区 Province	因旱饮水困难人口 / 万人 Population having difficulties accessing drinking water /10,000 persons	因旱饮水困难大牲畜 / 万头 Number of bigger-sized livestock having difficulties accessing drinking water/10,000 heads	地区 Province	因旱饮水困难人口 / 万人 Population having difficulties accessing drinking water /10,000 persons	因旱饮水困难大牲畜 / 万头 Number of bigger-sized livestock having difficulties accessing drinking water/10,000 heads
全国 Nationwide	542.38	331.92	河南 Henan	7.36	2.30
北京 Beijing			湖北 Hubei	85.39	9.28
天津 Tianjin			湖南 Hunan	46.70	12.80
河北 Hebei	0.16	0.36	广东 Guangdong	2.80	
山西 Shanxi	5.13	1.87	广西 Guangxi	0.87	0.09
内蒙古 Inner Mongolia	18.04	127.85	海南 Hainan		
辽宁 Liaoning			重庆 Chongqing	121.83	62.88
吉林 Jilin			四川 Sichuan	76.88	52.80
黑龙江 Heilongjiang			贵州 Guizhou	45.68	14.39
上海 Shanghai			云南 Yunnan	39.14	12.35
江苏 Jiangsu			西藏 Xizang		
浙江 Zhejiang	1.69	0.01	陕西 Shaanxi	27.23	7.70
安徽 Anhui	0.25	0.36	甘肃 Gansu	10.10	4.47
福建 Fujian	15.66	5.57	青海 Qinghai	33.72	13.79
江西 Jiangxi	1.97	0.06	宁夏 Ningxia	1.68	2.28
山东 Shandong			新疆 Xinjiang	0.10	0.71

图 4-1　2022 年全国干旱灾害分布图

Figure 4-1　Overview of drought disasters nationwide in 2022

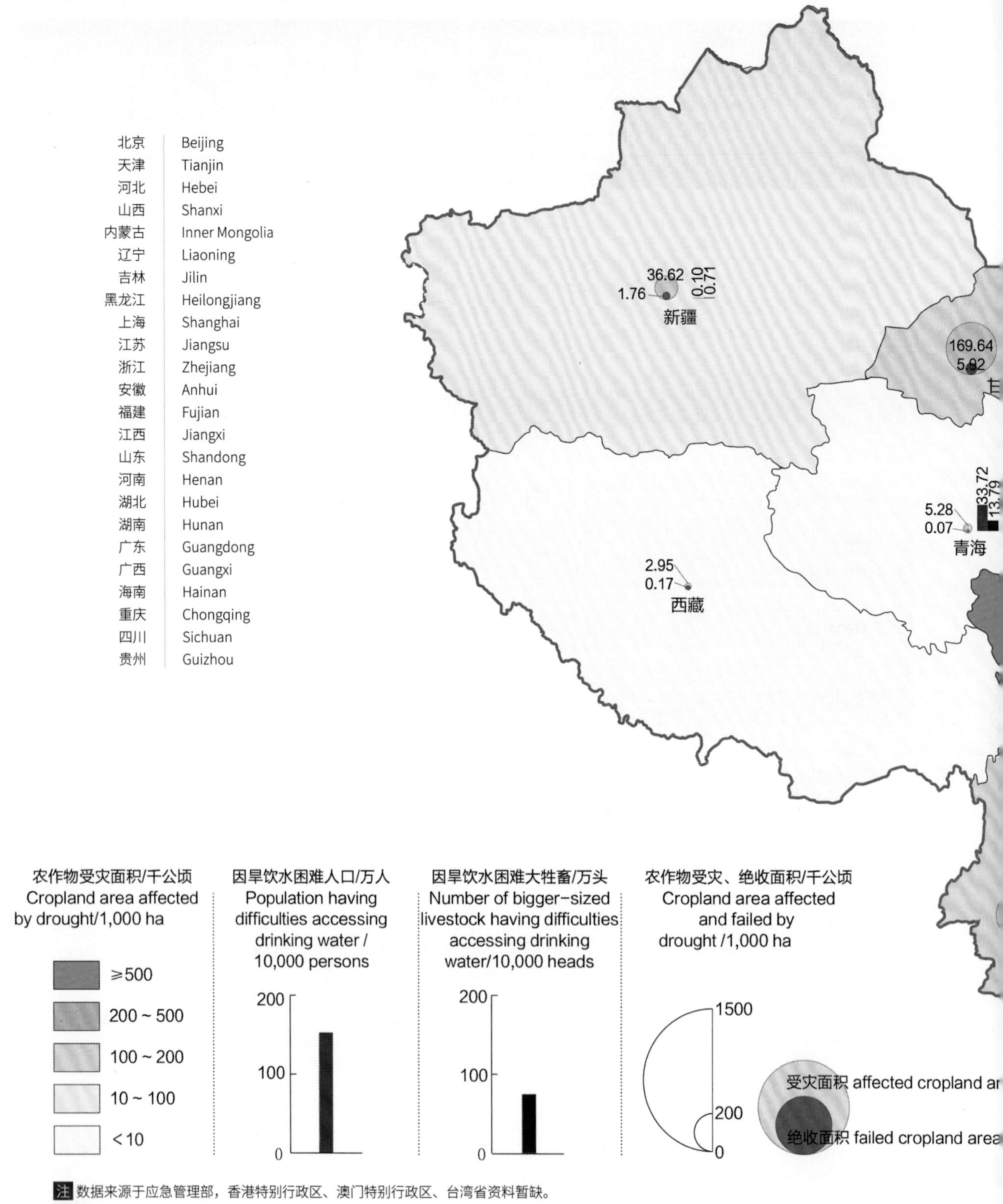

注 数据来源于应急管理部，香港特别行政区、澳门特别行政区、台湾省资料暂缺。

Note The data come from the Ministry of Emergency Managment , data of Hong Kong SAR, Macao SAR, and Taiwan are currently unavailable.

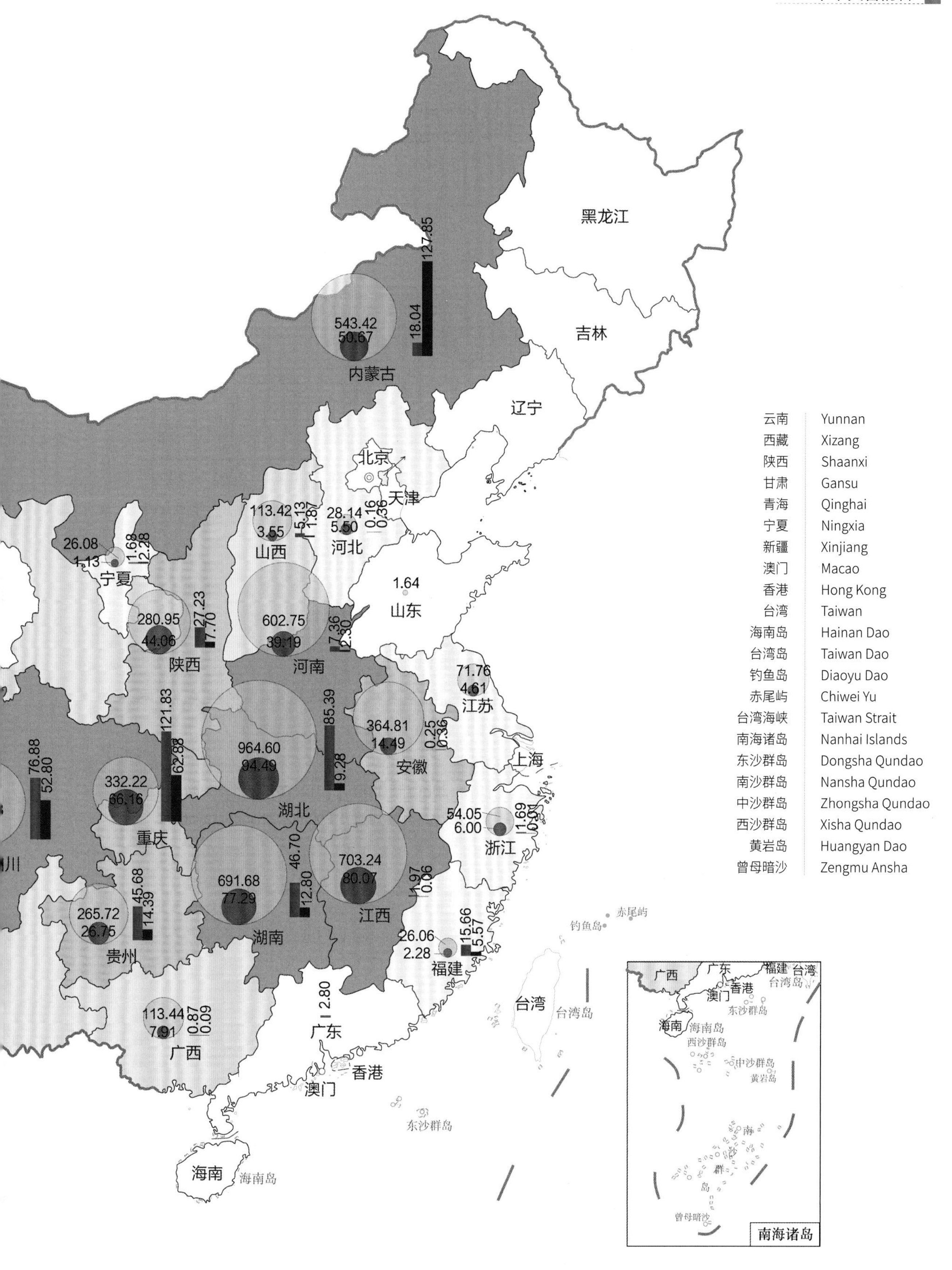

云南	Yunnan
西藏	Xizang
陕西	Shaanxi
甘肃	Gansu
青海	Qinghai
宁夏	Ningxia
新疆	Xinjiang
澳门	Macao
香港	Hong Kong
台湾	Taiwan
海南岛	Hainan Dao
台湾岛	Taiwan Dao
钓鱼岛	Diaoyu Dao
赤尾屿	Chiwei Yu
台湾海峡	Taiwan Strait
南海诸岛	Nanhai Islands
东沙群岛	Dongsha Qundao
南沙群岛	Nansha Qundao
中沙群岛	Zhongsha Qundao
西沙群岛	Xisha Qundao
黄岩岛	Huangyan Dao
曾母暗沙	Zengmu Ansha

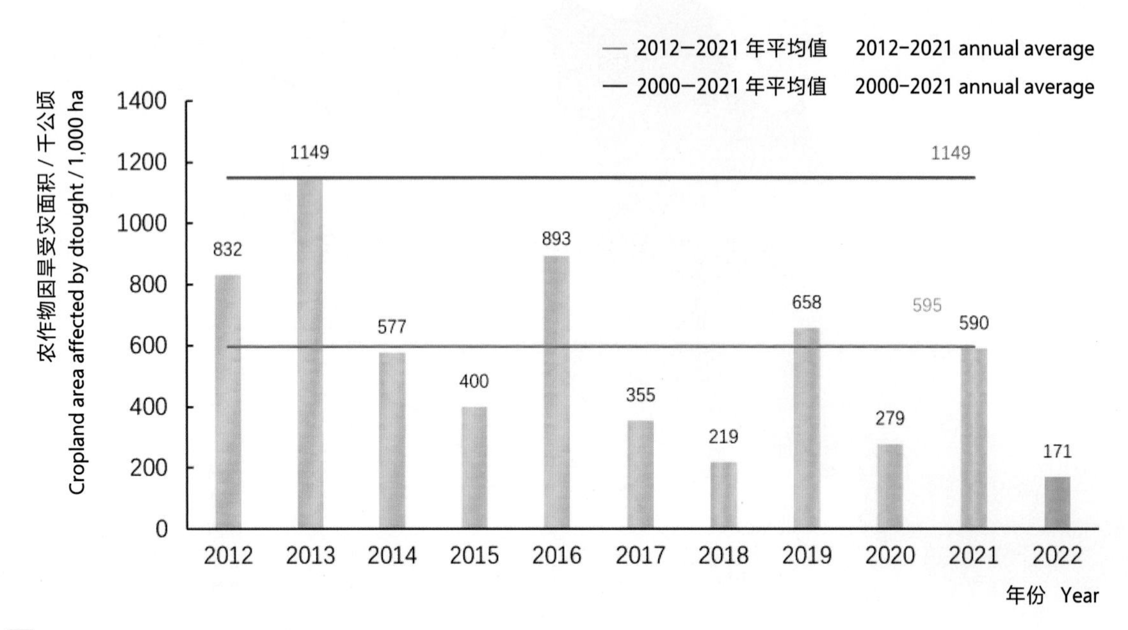

注 2019–2022 年数据来源于应急管理部。

Note The data during 2019-2022 come from the Ministry of Emergency Management.

图 4-2 2012—2022 年全国农作物因旱受灾面积统计

Figure 4-2 Cropland area affected by drought 2012–2022

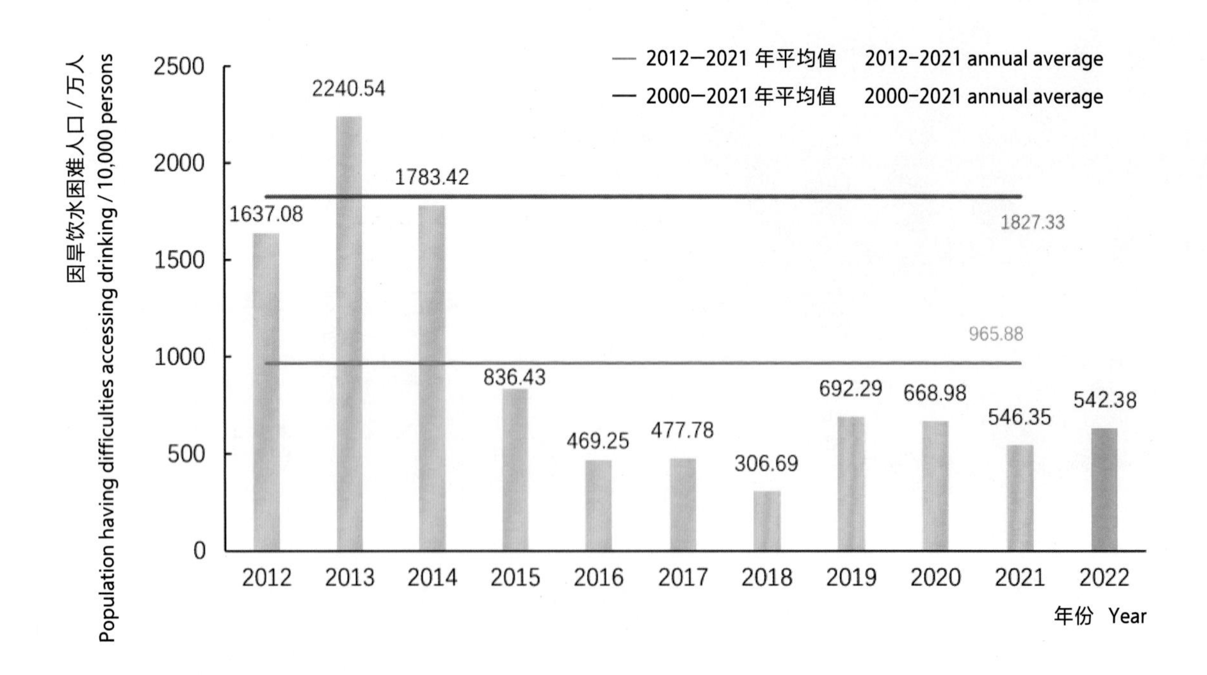

图 4-3 2012—2022 年全国因旱饮水困难人口统计

Figure 4-3 Population having difficulties accessing drinking water 2012–2022

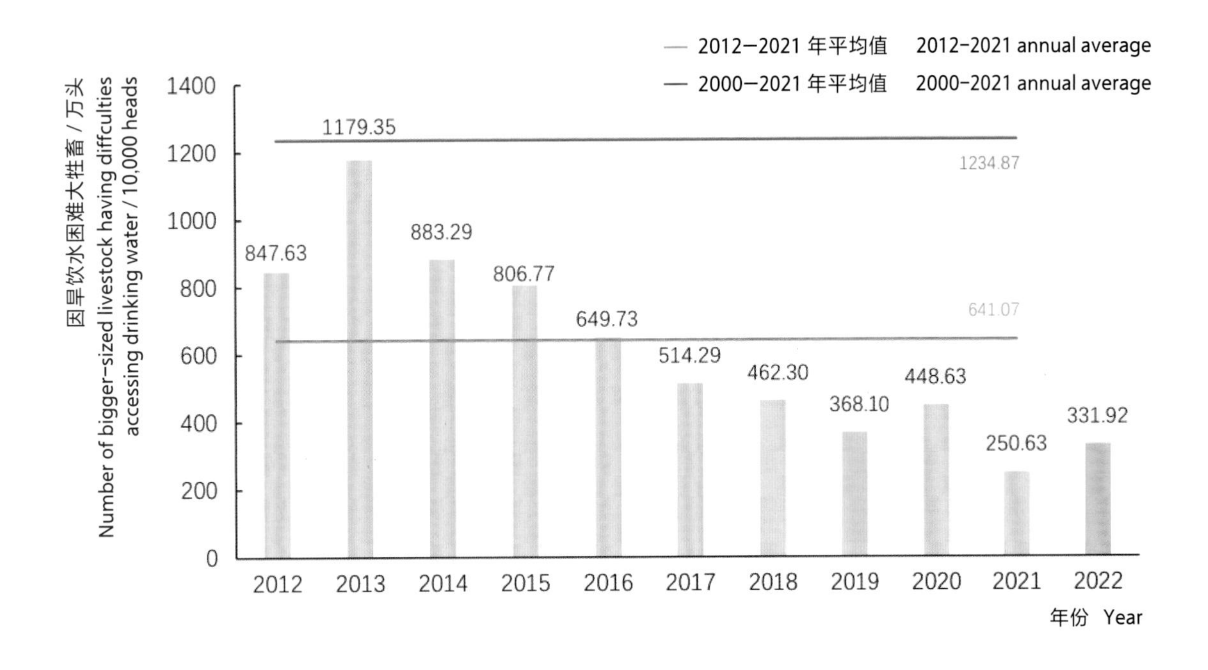

图 4-4 2012—2022 年全国因旱饮水困难大牲畜统计

Figure 4-4 Number of bigger-sized livestock having difficulties accessing drinking water 2012−2022

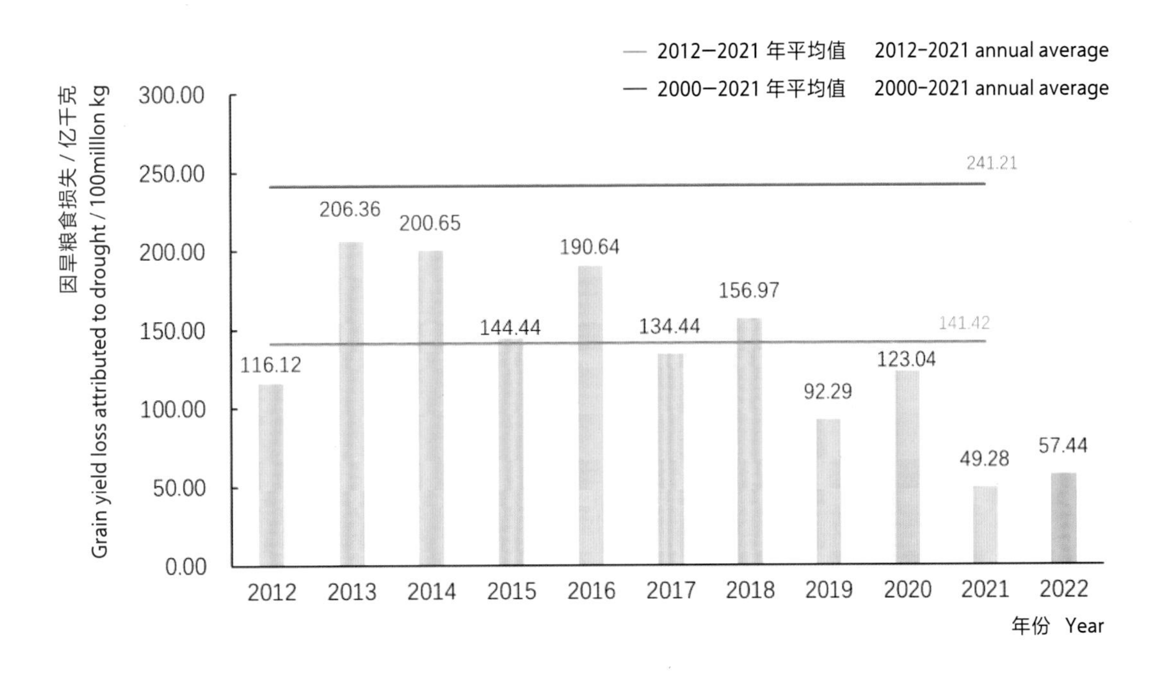

图 4-5 2012—2022 年全国因旱粮食损失统计

Figure 4-5 Grain yield loss attributed to drought 2012−2022

4.4 防御工作

水利部及地方各级水利部门认真贯彻落实习近平总书记关于防汛抗旱重要指示批示精神和党中央、国务院决策部署，坚持“预”字当先、“实”字托底，精准范围、精准对象、精准措施，有序做好各项抗旱工作。

4.4.1 工作部署

水利部部长李国英针对珠江流域西江、东江、韩江“冬春连旱、旱上加咸”态势，视频连线珠江委和广东、福建两省，提出“确保香港、澳门供水安全，确保珠江三角洲城乡居民生活用水安全”的目标，安排部署珠江流域压咸补淡保供水行动；针对长江流域夏秋连旱，多次召开抗旱专题会商会，视频连线长江委和相关省（直辖市），分析研判旱情形势，部署抗旱保供水保灌溉工作，并在抗旱关键时期赶赴重庆、湖北、湖南、江西等省（直辖市）旱区一线，与相关省（直辖市）领导共商抗旱对策，研究实施两轮长江流域水库群抗旱保供水联合调度专项行动；针对上海市供水受咸潮影响情况，召开抗咸潮保上海市供水专题会商会，视频连线长江委、太湖局和上海市，研究实施抗咸潮保供水专项行动，指导长江委、太湖局做好长江、太湖水源供水保障工作，全力确保上海市供水安全。分管部领导多次召开会商会或视频会，对抗旱工作作出具体安排，并赴广东、福建、湖北、江西等省调研指导抗旱工作。水利部先后发出 12 个通知，对旱情监测、水库调度、供水保障、水源管理和完善预案等提出具体要求。派出 16 个工作组赴旱区一线，协助指导地方因地制宜落实水库补水、应急调水、打井取水、架泵提水和加强节水等抗旱保供措施。

4.4 Prevention and Control

MWR and local water resources departments have implemented the instructions for flood control and drought relief made by President Xi Jinping and the decisions and employment of the CPC Central Committee and the State Council. The Ministry insists on prevention as the most important orientation, puts preparedness first based on concrete measures, precisely locates the disaster range and the community being protected with targeted measures, in a bid to fight and relieve drought disasters.

4.4.1 Arrangements

Minister Li Guoying, in response to the winter-spring stretching drought and salinity increase, convened several online meetings with Pearl River Commission and local authorities of Guangdong and Fujian provinces; raised the goals of "securing water supplies in Hong Kong and Macao and for residents along the Pearl Delta region", and deployed relevant actions for checking saltwater intrusion and guaranteeing water supply in the Pearl River basin. To combat summer-autumn drought in the Yangtze basin, Minister Li convened several consultation meetings at which the drought conditions were analyzed and assessed, and relevant deployments were made to assure water supply and irrigation. The Minister also visited the drought-stricken areas in Chongqing Municipality, Hubei Province, Hunan Provinces and Jiangxi Province at the critical moments and discussed with leaders of the jurisdictions on the countermeasures for drought mitigation; and implemented two rounds of special actions for drought relief through joint dispatch of reservoir groups in the Yellow River Basin. In response to saltwater intrusion in Shanghai, multiple consultation meetings with online participation of the Changjiang Commission, the Taihu Authority and Shanghai local authority were arranged, and special actions were carried out accordingly so as to guarantee the safety of water supply in the city. Leaders of the Ministry who are in charge made specific arrangements for drought mitigation through multiple consultation or online meetings, and visited Guangdong, Fujian, Hubei and Jiangxi provinces/autonomous regions to conduct study and guide drought relief. Twelve notices were issued by MWR stipulating drought monitoring, reservoir regulation, water supply support, water resources management and contingency planning. Sixteen working groups were sent by the Ministry to the front line of the drought-affected regions, helping guide the local efforts in implementing measures including reservoir replenishment, emergency water diversion plans, well drilling, water pumping and water saving reinforcement, etc.

长江委多次召开会议安排推进抗旱重点任务，派出 9 个工作组赴旱区指导地方做好抗旱保供水工作，派专家参加水利部抗咸潮保供水工作组，指导上海市应对咸潮入侵。黄委发出通知安排部署旱情监测预报预警、水库调度、抗旱措施落实等工作，派出工作组赴内蒙古自治区开展抗旱水源保障工程调度、设施设备维护等督导检查。淮委发出通知指导地方科学制定用水计划，保障农业、生活和生产用水需求，派出工作组实地调研旱情灾情，指导抗旱工作。珠江委派出 13 批次工作组，深入旱区和咸潮影响区域，了解供水需求，研判供水形势，指导基层落实各项措施。太湖局持续做好会商研判，滚动分析旱情和供水形势，强化与有关省（直辖市）的沟通联系，督促做好蓄水保水供水和水资源调度指导。

4.4.2 “四预”措施

水利部强化预报、预警、预演、预案“四预”措施，密切关注旱区雨情、水情、旱情，滚动预测预报，及时发布干旱预警，加强水量供需分析演算，指导旱区滚动完善抗旱预案，落实抗旱保供水兜底措施；长江流域夏秋连旱期间，多次发布枯水预警信息，根据中下游灌区秋粮作物用水需求，科学编制调度方案；在上海市抗咸潮专项行动期间，严密监测三峡水库补水径流演进、长江口咸潮上溯、长江沿线主要口门及闸站引水流量等信息，准确预报长江口补水最大径流与最小潮头对接时段，为调度决策提供有力支持；组织长江、黄河、海河、珠江等流域管理机构编制多个应急水量调度预案，为做好流域应急水量调度工作提供重要依据；加快推进全国旱情监测预警综合平台建设工作，为抗旱决策提供有力技术支撑；编制完成《江河湖库旱警水位（流量）计算方法案例》，组织各流域管理机构和各省级水行政主管部门开展 1200 多处水文测站旱警水位（流量）确定工作。

The Changjiang Commission held multiple meetings to facilitate the prioritized tasks of drought mitigation. Nine working groups were sent to drought areas to provide guidance for water supply; relevant experts from the Commission joined the Ministerial working group in fighting saltwater intrusion in Shanghai. The Yellow River Commission sent notices for deploying the drought monitoring, forecasting and early-warning, reservoir regulation, countermeasures for drought mitigation, etc.; and working groups were dispatched to Inner Mongolia for supervision and inspection of drought-resistant water supply project, facilities and equipment maintenance, etc. The Huaihe Commission provided the local governments with guidance on how to scientifically make water use plans in a bid to safeguard agricultural, household and production water needs; working groups were also sent to the drought-stricken areas for conducting field research and providing professional guidance. While the 13 batches of working groups appointed by the Pearl River Commission went to the regions affected by drought and saltwater intrusion to investigate the water demands, in a bid to assess the water supply situation and guide the communities to implement the measures. The Taihu Authority frequently held consultation meetings and analyzed the drought situations and water supply dynamically, enhanced the communications with relevant provinces/municipalities, supervised and guided water storage, water protection, water supply and water resources scheduling.

4.4.2 The "four preemptive pillars"

The MWR has shored up the "four preemptive pillars" - forecasting, early warning, exercising and contingency planning; closely tracked rainfall, water regime and drought conditions; conducted non-stop forecasting and prediction, and issued early warnings promptly; strengthened the analysis and calculation of water supply and demand, led the drought areas to improve their drought relief contingency plan constantly and implement fallback measures to safeguard water supply. During the summer-autumn drought in the Yangtze river basin, the Ministry issued multiple low-water warnings, and scientifically developed the water dispatch plan based on the water demand of autumn crops in irrigation districts in the middle and lower reaches. In response to the saltwater intrusion in Shanghai, MWR strictly monitored the key information including the runoff routing replenished by the Three Gorges Reservoir, the saltwater intrusion in the Yangtze estuary, the diverted water flow of the main gates and stations along the Yangtze River, so as to accurately forecast the proper time period when reaching the maximum runoff and the lowest salt tide, thus providing reliable support to the scheduling decisions. The Ministry also organized the river basin authorities of the Yangtze, the Yellow, the Haihe and the Pearl rivers to make several contingency plans for water dispatch as an important basis for emergent water dispatch. While a comprehensive and nationwide platform for drought monitoring and forecasting has been pushed forward, serving as a strong technical support for decision making of drought prevention. *The Case Studies on Calculation Methods of Drought-Alert Levels (Flows) In Rivers, Lakes, and Reservoirs* was drafted, and the main river basin commissions and provincial water authorities were all coordinated to determine the drought-alert levels (flows) at more than 1200 stations.

长江委预测研判流域可能发生严重干旱，提醒指导流域各地提前开展水库蓄水保水和沿江引水保水，针对长江中下游抗旱补水和长江口抗咸潮补水滚动编制更新水库群联合调度方案，支撑调度决策；汛后及时审查批复控制性水库蓄水计划，为后期抗旱有效储备水源。珠江委依托珠江流域抗旱“四预”平台，动态预演供水保障“三道防线”，研究制定“最不利情况下坚守底线”的调度方案，为抗旱工作赢得主动。太湖局在实施引江济太应急调水期间，每日滚动开展预报预演，模拟长江水沿望虞河运动变化过程，引水入湖后立即组织反演，检验预演效果，同时加强区域降雨情况下望虞河水位变化预演，严防旱涝急转。上海市开展水源地咸潮入侵滚动预报，密切关注台风、冷空气对取水口引水的影响，做好跟踪研判。江西省实施“三个 10 天”旱情预警机制，滚动预报未来 10 天、20 天、30 天的旱情发展态势。湖北省加密蒸发量、土壤墒情、枯水水位等信息监测，墒情站由 10 日一测报改为 5 日一测报，蒸发站每日报送数据。湖南省切实落实 7 天滚动预测、5 天研判预警、3 天调度交办、1 天督促落实的“7531”抗旱工作机制。重庆市对全市受干旱影响的农村人口供水保障情况持续开展日监测、日巡查、日调度，动态监测水源水量变化。

Upon predicting that there might be severe droughts after study, the Changjiang Commission urged preemptive impoundment by reservoirs and water diversion along the Yangtze river to secure the water supply; continuously updated the joint operation plans of reservoirs in the middle and lower reaches to fight the drought and check saltwater intrusion in the estuary; the water storage plans of controlling reservoirs were reviewed after the flooding responsively, thereby providing available water resources for future drought prevention. The Pearl River Commission, drawing upon the digital platform of forecasting, early warning, exercising and contingency planning, rehearsed dynamically the "three safety nets" of water supply and proposed the water dispatch plans for "upholding the bottom line even facing the worst scenario" in a proactive manner. When the Taihu Authority was implementing the emergency water diversion of Yangtze-Taihu, forecasting and exercising were conducted on a daily basis in simulating the movement of the Yangtze water routing along the Wangyu River, and reflective demonstration was organized immediately after the diversion for evaluating the performance of previous exercise; in the meanwhile the rehearsal of the water level change in the Wangyu River under the condition of regional rainfall was enhanced to prevent abrupt drought-flood transition. Shanghai carried out constant forecast of saltwater intrusion in the source water areas, closely monitored the impact of typhoons and cold air on water intakes for diversion, thus ensuring sound follow-up and judgement. A drought early-warning mechanism called "three 10 days" was applied in Jiangxi province, namely, continuously forecast the trend of drought conditions for the next 10, 20 and 30 days. Hubei province increased the monitoring frequency on information including evaporation, soil moisture, low water level; specifically, the soil moisture station was operated from every 10 days to every 5 days, and the evaporation monitoring station generated data on a daily basis. Hunan province stuck to a "7531" drought-resistant work routine of 7-day non-stop forecasting, 5-day study and early warning, 3-day dispatch and arrangement and 1-day supervision and implementation. Chongqing Municipality undertook a day-by-day monitoring, inspection and dispatch on the water supply for drought-affected rural population, and dynamically tracked the change in water amount of the water sources.

4.4.3 启动响应

面对严峻旱情，水利部启动干旱防御Ⅳ级应急响应 3 次，累计 98 天。长江委启动干旱防御Ⅳ级应急响应 2 次，累计 81 天。黄委、淮委、太湖局均启动干旱防御 Ⅳ 级应急响应 1 次。长江流域旱区各省（直辖市）均启动了 Ⅳ 级及以上等级的干旱防御或抗旱应急响应，其中江西省先后启动了 Ⅲ 级和 Ⅱ 级应急响应，湖北、湖南、重庆、四川 4 省（直辖市）启动了 Ⅲ 级应急响应。

表 4-3 2022 年水利部本级和流域管理机构干旱防御应急响应启动情况
Table 4-3 Emergency responses against drought launched by MWR and river basin commissions in 2022

启动部门 Agency	启动时间 Start date	响应级别 Responses level	响应范围 Areas	终止时间（终止范围） End date (Areas of response)	持续时间 / 天 Duration/day
水利部 The Ministry of Water Resources	6 月 25 日 June 25	Ⅳ	内蒙古、河南、陕西、甘肃 Inner Mongolia, Henan, Shaanxi, Gansu	7 月 14 日 内蒙古、河南、陕西、甘肃 July 14 (Inner Mongolia, Henan, Shaanxi, Gansu)	20
	8 月 11 日 August 11	Ⅳ	安徽、江西、湖北、湖南、重庆、四川 Anhui, Jiangxi, Hubei, Hunan, Chongqing, Sichuan	9 月 7 日 四川 September 7 (Sichuan)	28
				10 月 27 日 安徽、江西、湖北、湖南、重庆 October 27 (Anhui, Jiangxi, Hubei, Hunan, Chongqing)	78
	8 月 22 日 August 22	Ⅳ	江苏、河南、贵州、陕西 Jiangsu, Henan, Guizhou, Shaanxi	9 月 7 日 江苏、河南、陕西 September 7 (Jiangsu, Henan, Shaanxi)	16
				10 月 27 日 贵州 October 27 (Guizhou)	66

续表 Continued

启动部门 Agency	启动时间 Start date	响应级别 Responses level	响应范围 Areas	终止时间（终止范围） End date (Areas of response)	持续时间 / 天 Duration/day
长江委 The Changjiang River Commission	8月11日 August 11	Ⅳ	四川、重庆、湖北、湖南、江西、安徽 Sichuan, Chongqing, Hubei, Hunan, Jiangxi, Anhui	9月17日 四川 September 17 (Sichuan)	38
				10月11日 湖北 October 11 (Hubei)	62
				10月21日 安徽 October 21 (Anhui)	72
				10月30日 重庆、湖南、江西 October 30 (Chongqing, Hunan, Jiangxi)	81
	8月22日 August 22	Ⅳ	江苏、贵州、陕西、河南 Jiangsu, Guizhou, Shaanxi, Henan	9月1日 江苏、陕西、河南 September 1 (Jiangsu, Shaanxi, Henan)	11
				10月11日 贵州 October 11 (Guizhou)	51
黄委 The Yellow River Commission	6月25日 June 25	Ⅳ	甘肃、内蒙古、陕西 Gansu, Inner Mongolia, Shaanxi	7月14日 甘肃、内蒙古、陕西 July 14 (Gansu, Inner Mongolia, Shaanxi)	20
淮委 The Huaihe River Commission	8月22日 August 22	Ⅳ	淮南地区、沿淮地区 Northern and riverine regions of the Huaihe River	8月27日 淮南地区、沿淮地区 August 27 (Northern and riverine regions of Huaihe River)	6
珠江委 The Pearl River Commission	2021年10月16日 October 16 2021	Ⅳ	珠江流域 The Pearl River basin	2022年3月24日 珠江流域 March 24 (The Pearl River basin)	160
太湖局 The Taihu Lake Authority	8月23日 August 23	Ⅳ	太湖流域片 The Taihu Lake basin	8月28日 太湖流域片 August 28 (The Taihu Lake basin)	6

4.4.3 Emergency responses

In face of harsh drought conditions, MWR launched level Ⅳ emergency responses against drought for three times, lasting 98 days. The same level of emergency responses were launched twice by the Changjiang Commission for 81 day in total, and once by the Yellow River Commission, the Huaihe Commission and the Taihu Authority, respectively. The drought-stricken provinces/municipalities along the Yangtze basin all launched level Ⅳ or above emergency responses against drought, among which Jiangxi province launched level Ⅲ and level Ⅱ emergency responses, while Hubei, Hunan, Chongqing and Sichuan provinces/municipality launched level Ⅲ emergency responses.

4.4.4 应急调水

水利部指导有关流域管理机构和地方水利部门，精细调度水工程，全力应对旱情。针对珠江流域旱情，珠江委构筑当地、近地、远地供水保障“三道防线”，先后 3 次启动压咸补淡应急调度，联合调度珠江流域天生桥、龙滩、百色、大藤峡、新丰江、枫树坝等骨干水库，保障了香港、澳门和珠江三角洲城市供水安全。针对黄淮海和西北地区旱情，黄委精细调度黄河龙羊峡、刘家峡、万家寨、小浪底等骨干水利工程，保障了下游灌区灌溉用水；指导地方根据不同作物时令需水情况，因时因地施策，科学制定灌溉用水计划，合理配置灌溉水量。在长江流域秋粮作物生长关键时段，水利部组织长江委和江西、湖南两省水利厅实施两轮长江流域水库群抗旱保供水联合调度专项行动，精准调度长江流域 75 座大中型水库，累计补水 61.6 亿立方米，指导下游精准对接每一个灌区、每一个取水口，优化调整灌溉计划，保障了补水沿线 356 处大中型灌区及众多小型灌区用水。针对长江口咸潮上溯影响上海市城市供水的严峻局面，水利部组织长江委、太湖局和上海市水务局实施上海市抗咸潮保供水专项行动，调度三峡水库向下游补水 40.6 亿立方米，中下游沿程采取引水管控措施，有效压制咸潮；及时增加太浦河供水流量，加大引江济太力度，快速打通太湖和望虞河向陈行水库应急补水通道，保障了上海 90% 自来水原水供应，确保上海市供水安全。针对太湖流域旱情，实施望虞河、新孟河双通道引水，全年引长江水 27.1 亿立方米，直接入太湖 11.9 亿立方米，向湖西区运河南部区域补水 3.9 亿立方米，有效增加了流域水量。针对淮河流域洪泽湖周边旱情，指导调度沂沭泗河水系向洪泽湖周边补水近 15 亿立方米；实施引沂济淮，有效缓解里下河地区旱情，保证苏北地区城乡供水安全和粮食丰收。

江苏省通过江水北调、江水东引等工程累计抽引长江水约 35 亿立方米，2200 千公顷水稻未受到干旱影响。江西省调度峡江、廖坊等 36 座大中型水库为下游补水 24.5 亿立方米，保障了 433.33 千公顷农田用水需求。湖北省调度引江济汉工程从长江向汉江、长湖应急补水 20.9 亿立方米，调度鄂北水资源配置工程向襄阳、随州、孝感等地补水 2 亿立方米。湖南省调度五强溪、柘溪等 35 座大中型水库，累计为下游补水 50 多亿立方米，调动洞庭湖北部补水工程等向洞庭湖区补水 23.5 亿立方米。重庆市做好水库、泵站、水闸等水工程调度，对接 1800 余处城乡取水口，累计调水超 3 亿立方米。

4.4.4 Emergency water dispatch

Led by MWR, the river basin commissions and local water authorities accurately dispatched the water projects and grappled with the challenges brought by droughts. The Pearl River Commission managed to build the "three safety nets" of local, nearby, and long-distance source water allocation, initiated emergency freshwater dispatch against saltwater intrusion for three times, and joint regulations were implemented on the operation of several major reservoirs including Tianshengqiao, Longtan, Baise, Datengxia, Xinfengjiang, Fengshuba, thereby securing the water supply in Hong Kong, Macao, and the Pearl Delta cities. To combat the drought in Huang-Huai-Hai region and Northwest China, the Yellow River Commission accurately commanded the operations of several backbone water projects including Longyangxia, Liujiaxia, Wanjiazhai and Xiaolangdi, ensuring the irrigation water use in the lower reaches; and guided the local departments to take targeted measures tailored to local conditions and times according to the seasonal water demand of crops, and make rational water use plans for irrigation, so as to reasonably allocate the irrigation water. During the critical growth period of autumn crops in the Yangtze basin, the Ministry mobilized the Changjiang Commission, along with the provincial water resources departments of Jiangxi and Hunan, to put dedicated efforts in two drought-resistance and water supply special actions through joint dispatch of reservoir groups, regulating 75 large and medium-sized reservoirs in the Yangtze basin with high precision, replenishing 6.16 billion m^3 of water in total; and gave instructions to the lower reaches on connecting every irrigation area and water intake to improve the irrigation plans, thereby safeguarding the water use in 356 large and medium-sized irrigation areas and multiple small irrigation areas along the replenishment route. To cope with the challenge posed by saltwater intrusion to urban water supply in Shanghai, led by MWR, the Changjiang Commission, the Taihu Authority and Shanghai Municipal Water Resources Department implemented special actions for preventing saltwater intrusion and securing water supply; specifically, by releasing water from the Three Gorges reservoir with an amount of 4.06 billion m^3 downstream, taking water diversion and control measures along the middle and lower reaches to effectively curb the entry of saltwater, enlarging the discharge of water supplied by the Taipu River, reinforcing the Yangtze-Taihu water diversion project, promptly enabling the emergency replenishment channel from Taihu Lake and Wangyu River to the Chenhang Reservoir, thereby securing 90% tap and raw water supply in Shanghai. In response to the drought in the Taihu Lake basin, dual channel of the Wangyu River and the Xinmeng River was available to divert 2.71 billion m^3 of water from the Yangtze River, among which 1.19 billion m^3 of water directly flowed into the Taihu Lake, and 390 million m^3 of water was diverted to the southern region of the west lake, effectively

4.4.5 抗旱投入

党中央、国务院对抗旱工作高度重视，国务院常务会研究安排 100 亿元中央预备费支持以长江流域为重点的受旱地区抗旱减灾，其中 65 亿元用于支持水利抗旱救灾工作。水利部商财政部安排中央水利救灾资金 2 亿元，支持黄淮海和西北地区开展抗旱打井、建设蓄引提调等抗旱应急工程、添置提水运水设备和补助抗旱用油用电；召开视频会要求用好中央救灾资金、加大地方抗旱投入、加强资金监管、优化建设流程、加快项目实施进度，尽快将资金转化为抗旱能力。2022 年全国共投入抗旱劳力 1928.18 万人，开动机电井 137.77 万眼、泵站 18.86 万处、机动抗旱设备 300.49 万台（套），出动各类机动运水车辆 28.95 万辆，各级财政累计投入抗旱资金 156.42 亿元。

河北省安排特大抗旱补助资金 1000 万元，支持 32 个县（市、区）抗旱应急工程建设、添置抗旱应急设施，建设一批抗旱应急工程，在夏粮丰收中发挥了重要作用。福建漳州市投入 3000 万元，改造云霄县城提水设施，保障供水安全；龙岩市加快推进万安溪引水工程，保障主城区供水安全。湖北省新建管网 2076 千米、新建加压站 174 处、打机电井 313 处、筑拦河坝 96 处、新建水源 991 处，解决群众饮水困难问题。湖南省利用地下水位浅、水量大的优势，累计打井 11000 余口，发挥地下水抗旱战略储备水源作用；启用应急水源洞庭湖北部补水工程，保障 80 多万人生活用水需求。广东省建成潮州引韩济饶、梅州丰顺新区、揭阳普宁北部中心等 3 处重点抗旱保供水工程，完善抗旱工程体系。重庆市启用渝西水资源配置工程中急用先建项目，运用长江上中游最大规模趸船应急取水，缓解江津以北地区 70 万群众用水紧缺局面。

江西吉安市村民打井疏渠抗旱（8 月 18 日）
Villagers in Ji'an, Jiangxi province was digging and dredging to fight drought (August 18)

increasing the water amount of the Taihu basin. Aiming at the drought nearby the Hongze Lake of the Huaihe basin, about 1.5 billion m^3 of water from the Yihe-Shuhe-Sihe River System was released to the Hongze Lake region, and the Yihe-Huaihe water diversion was implemented in a bid to mitigate the drought conditions over the lower reaches and ensure urban and rural water supply and agriculture in the northern Jiangsu.

Though water diversion from the Yangtze River to its northern and eastern parts, Jiangsu province drew approximately 3.5 billion m^3 of water from the Yangtze River to protect around 2,200,000 ha of rice crops from drought. Jiangxi province dispatched 2.45 billion m^3 of water from 36 large and medium-sized reservoirs to the downstream, ensuring the water demand of 433,330 ha of cropland. Hubei province, relying on the Yangtze-Hanjiang water diversion project, replenished 2.09 billion m^3 of water to the Hanjiang River and the Changhu Lake, and replenished 200 million m^3 of water using water resources allocation projects in the north Hubei to Xiangyang, Suizhou, Xiaogan and other regions. Hunan province dispatched 35 large and medium-sized reservoirs such as Wuqiangxi and Zhexi reservoirs, recharging the downstream at an amount of 5 billion m^3 water in total, and replenishing 2.35 billion m^3 of water to the Dongting Lake district by operating the water replenishment engineering projects in northern area of the lake. Chongqing was well prepared for scheduling water projects including reservoirs, pumping stations and gates, and checked over 1800 water intakes in urban and rural areas. More than 300 million m^3 of water was diverted.

4.4.5 Financial and in-kind input for drought relief

The CPC Central Committee and the State Council attached great importance to drought prevention. At the executive meeting of the State Council, it was approved that 10 billion RMB of the central reserve fund was used for supporting drought mitigation in disaster-affected areas with the Yangtze basin as a priority, among which 6.5 billion RMB was used for supporting drought prevention and mitigation in the water conservancy sector. MWR, in consultation with the Ministry of Finance, allocated 200 million RMB of the central water disaster relief fund to support the Yellow River and Huaihe basins as well as northwest China in wells digging, constructing drought-resistant emergency projects for water storage, extraction and diversion, purchasing equipment for water extraction and transportation, and subsidizing the cost of engine oil and electricity consumed for drought prevention. Video meetings were organized to ensure the good and practical use of the central disaster relief fund, increase investment in local drought-resistance actions, step up efforts in financial supervision, accelerate construction workflow, in a bid to speed up the progress of relevant projects. In 2022, a total of 19,260,300 people participated in the drought-resistance

安徽当涂县抽取芜申运河河水进入内河抗旱（8 月 18 日）
Water from the Wushen canal was abstracted to the river in Dangtu, Anhui province for drought prevention (August 18)

4.5 防御成效

2022 年，面对历史罕见的严重干旱，各级水利部门坚持人民至上、生命至上，牢牢扛起抗旱天职，累计解决饮水困难人口 520.52 万人，完成抗旱浇地面积 14267.71 千公顷，挽回粮食损失 156.71 亿千克、经济作物损失 228.79 亿元，实现了“确保旱区群众饮水安全、保障大牲畜饮水和秋粮作物时令灌溉用水需求”的既定目标，同时有效保障了洞庭湖、鄱阳湖、太湖等水生态环境安全。

actions as labor force, 1,376,700 electromechanical wells, 188,500 pumping stations and 3,004,100 sets of mobile equipment were put into operation, and 289,500 water tankers were commissioned. The overall investment in drought prevention by all levels of financial departments reached 15.634 billion RMB.

Hebei province arranged 10 million RMB from the provincial fiscal coffers as a large-scale drought relief fund, intending to support 32 towns/cities/districts in constructing drought-resistance emergency projects and purchasing emergency facilities that played key roles in summer crops production. Zhangzhou city in Fujian province spent 30 million RMB to upgrade water extraction facilities in Yunxiao town to secure water supply; Longyan boosted the development of Wan'anxi Water Diversion Project for safeguarding water supply in the city proper. Hubei province newly built 2,076 km of water pipelines network, 174 booster pumping stations, 313 electromechanical wells, 96 weirs, and 991 water sources, so as to ease the pressure of drinking water supply for public. Hunan province, taking advantage of its shallow water tables with plentiful water resources, dug over 11,000 wells and gave full play to the groundwater as a strategic reserve of water resources during drought mitigation; and the north water recharge project of the Dongting Lake was put into operation, protecting more than 800,000 people's water needs. Guangdong province completed three major drought-resistant and water supply protection projects in Chaozhou (Hanjiang-Raoping Water Diversion Project), Meizhou (Fengshun new district) and Jieyang (northern center of Puning), which became critical part of the drought-resistant engineering system. Chongqing activated the first built projects of the Yuxi Water Resources Allocation Project, and the largest barge in the upper and middle reaches of the Yangtze River was applied to carry out emergency water collection, in an effort to alleviate the water shortage that affected 700,000 people in the north of Jiangjin.

4.5 Effectiveness of Drought Disaster Prevention

In 2022, encountering the historical extreme droughts, water resources departments at all levels insisted on the supremacy of the people and life, shouldered their duties of fighting drought disasters. Water was supplied to 5.2052 million people with temporary difficulties accessing drinking water, and a total of 14,267,710 ha of affected cropland were saved with emergency irrigation. As a result, 15.671 billion kg of grain yield and cash crops worth of 22.879 billion RMB were recovered. The stated goal of "ensuring drinking water security for the drought-stricken people, guaranteeing the drinking water needs of large livestock and the seasonal irrigation water use of autumn crops" was achieved, while water ecology of the Dongting, Poyang and Taihu lakes was well protected.

案例 1 两轮长江流域水库群抗旱保供水联合调度专项行动

在长江流域秋粮作物生长关键时段，水利部组织长江委实施了两轮长江流域水库群抗旱保供水联合调度专项行动。

8 月 16 日，启动第一轮专项行动。其间，调度以三峡水库为核心的长江上游干流梯级水库群补水 8.3 亿立方米，洞庭湖及湘、资、沅、澧“四水”，鄱阳湖及赣、抚、信、饶、修“五河”干支流水库群补水 27.4 亿立方米，使长江中下游沙市、七里山、汉口、湖口站水位较不补水情况抬高 0.40 ～ 0.10 米，改善了中下游沿江取水口的取水条件。长江中下游沿线湖北、湖南、江西、安徽、江苏 5 省共引水超过 26 亿立方米，农村供水受益人口 1385 万人，保障了 356 处大中型灌区 1904 千公顷农田用水需求。

9 月 12 日，启动第二轮专项行动。其间，以三峡水库为核心的长江梯级水库群补水 6.8 亿立方米，洞庭湖及湘、资、沅、澧“四水”和鄱阳湖及赣、抚、信、饶、修“五河”干支流水库群补水 19.1 亿立方米，使长江中下游沙市、七里山、汉口、湖口站水位较不补水情况下抬高 1.00 ～ 0.30 米，重点保障了长江中下游 973.33 千公顷秋粮作物灌溉用水需求。

结合专项行动，充分发挥三峡等骨干电站调峰调频作用，在增加中下游沿线水源保障的同时，有效保障有关地区高温时段的电力供应。

Case 1 Two rounds of drought relief and water supply special actions in the Yangtze River basin through joint dispatch of reservoir groups

During the critical growth period of autumn crops in the Yangtze basin, MWR mobilized the Changjiang Commission to put dedicated efforts in two drought relief and water supply campaigns through joint dispatch of reservoir groups.

The first round of actions was launched on August 16. The Yangtze River upstream cascade reservoir group with the Three Gorges as its core were dispatched and released 830 million m^3 of water. A total 2.74 billion m^3 of water was provided by the reservoir groups from the Dongting Lake, Xiangjiang River, Zishui River, Yuanjiang River, Lishui River, Poyang Lake, Ganjiang River, Fuhe River, Xinjiang River, Raojiang River and Xiushui River. The stations located in the middle and lower reaches of the Yangtze River including Yousha, Qilishan, Hankou and Hukou measured water levels 0.40-0.10 m higher as compared to the condition without replenishment, and the water extraction through the intakes along the middle and lower reaches became more accessible. Five provinces seated along the middle and lower reaches including Hubei, Hunan, Jiangxi, Anhui and Jiangsu diverted over 2.6 billion m^3 water, benefiting 13.85 million rural residents and securing the water demand of 1,904,000 ha of cropland in 356 large and medium-sized irrigation districts.

The second round of actions was launched on September 12. The Yangtze River upstream cascade reservoir group with the Three Gorges as its core released 680 million m^3 of water. A total 1.91 billion m^3 of water was provided by the reservoir groups from the Dongting Lake, Xiangjiang River, Zishui River, Yuanjiang River, Lishui River, Poyang Lake, Ganjiang River, Fuhe River, Xinjiang River, Raojiang River and Xiushui River. The stations located in the middle and lower reaches of the Yangtze River including Yousha, Qilishan, Hankou and Hukou measured water levels 1.00-0.30 m higher as compared to the condition without replenishment, thereby ensuring the irrigation water use of 973,330 ha of autumn crops downstream.

Relying on the special actions, the major hydropower stations such as the Three Gorges played a significant role in peak load and frequency regulation, thus effectively guaranteeing both the water sources along the middle and lower reaches and the electricity supply during high temperature.

案例 2 抗咸潮保上海市供水专项行动

受夏秋长江来水严重偏枯、东海台风盛行等因素影响，9 月上旬长江口咸潮上溯加剧，青草沙、陈行和东风西沙 3 座水库引水补库出现困难，上海市供水安全受到严重威胁。水利部持续密切关注长江口咸潮及其对上海市供水影响，部长李国英 9 月 27 日主持召开专题会商会，视频连线上海市政府、长江委、太湖局，共同研判上海市供水形势，提出“确保上海市生活、生产、生态用水安全；确保上海市社会大局稳定”的工作目标，明确“不能出现自来水限时供水，不能出现自来水降低水压，不能让市民喝咸水且出厂水质必须达标，不能号召市民储备和购买矿泉水，不能出现谣言四起、恶意炒作”的工作要求，启动实施抗咸潮保上海市供水专项行动。

水利部指导长江委紧急编制调度方案，在三峡水库来水、蓄水均严重偏枯的情况下，10 月 2—11 日启动三峡水库补水调度，并根据实时降雨、来水和咸情预报，将三峡水库泄量由 9000 立方米每秒加大至 12500 立方米每秒，累计向下游补水 40.6 亿立方米；结合缓解汉江下游船舶滞航需求，阶段性调度丹江口水库增加 500 立方米每秒下泄流量，并协调江苏、安徽两省 10 月 15—19 日压减沿江引水流量。太湖局加大引江济太力度，迅速打通太湖—河网—水库供水通道，保证陈行水库供水。上海市采取临时扩大引水口门、加设移动泵站等措施，精准捕捉引水窗口期并及时启动抽水补库。10 月 19—30 日青草沙、陈行和东风西沙 3 座水库共取水 5010 万立方米，增加了可供水量，有效缓解了上海市用水紧张局面。

Case 2 Special actions for resisting saltwater intrusion and securing water supply in Shanghai

Due to leaner-than-normal river inflow during summer and autumn as well as typhoons in the East China Sea, saltwater intrusion was aggravated in the Yangtze Estuary in early September. Three reservoirs including Qingcaosha, Chenhang and Dongfengxisha had difficulties diverting and replenishing water, considerably threatening the water supply in Shanghai. MWR paid close attention to the situation of saltwater intrusion in the Yangtze estuary and its impact on the city's water supply. Minister Li Guoying presided over a consultation meeting where he connected with Shanghai local government, the Changjiang Commission, the Taihu Basin Authority online and deliberated on the situation. He stated the goals of "safeguarding the domestic, production and ecological water use in Shanghai as well as the overall social stability"; and clearly required that efforts must be made to "avoid lowering water pressure over the taps, ensure quality drinking water, not incite the public to stock bottled water, and avoid rumors or malicious speculation". Afterwards the special actions for resisting saltwater intrusion and securing water supply in Shanghai were taken into effect.

In face of less-than-normal inflow and storage in the Three Gorges reservoir, the Ministry guided the Changjiang Commission to formulate emergency dispatch plans accordingly, and started to release water from the Three Gorges from October 2 to 11. Based on the real-time forecasting of rainfall, inflow and saltwater intrusion, the discharge of the Three Gorges was increased from 8,600 m^3/s to 12,500 m^3/s, providing 4.06 billion m^3 water to the downstream. Considering the demand for heaving the vessels in the lower Han River, the discharge flow of the Danjiangkou Reservoir was increased to 500 m^3/s as a phased approach, and Jiangsu and Anhui provinces were requested to reduce the water diversion discharge flow along the Han River. The Taihu Basin Authority stepped up the Yangtze-Taihu water diversion, promptly opened up the water supply channel among the Taihu, river network and reservoirs, so as to secure the water supply from the Chenhang Reservoir. Shanghai took temporary measures such as expanding the water intake gates, setting up movable pumping stations, etc., managing to seizing the best timing for water diverting and pumping. During October 19-30, 50.1 million m^3 water was drawn from three reservoirs including Qingcaosha, Chenhang and Dongfengxisha, thereby increasing the water supply and easing the water stress in Shanghai.

5 基础工作

FOUNDATIONAL WORK

5.1 机构职能

水利部优化调整蓄滞洪区建设管理有关职能，在水旱灾害防御司设立蓄滞洪区建管处，主要承担国家蓄滞洪区项目建设、日常监管、制度建设等职责。根据水利部党组决定，中国水利水电科学研究院优化调整有关内设机构设置，进一步强化水旱灾害防御支撑职能。

5.1 Institutional Functions

MWR has further clarified and divided the functions and responsibilities for constructing and managing the flood detention and retention basins. The Division of Flood Detention and Retention Basins Construction and Management was established accordingly under the Department of Flood and Drought Disaster Prevention of MWR, mainly in charge of project construction, routine supervision, institutional building related to flood detention and retention basins. China Institute of Water Resources and Hydropower Research, approved by the MWR Party group, has modified the organizational setup internally to better perform its support function in flood and drought disaster prevention.

5.2 规章制度

水利部修订印发《水利部水旱灾害防御应急响应工作规程》《关于加强山洪灾害防御工作的指导意见》。长江委修订印发《长江流域水旱灾害防御应急预案》《长江防汛简明手册》《长江委水旱灾害防御工作组、专家组管理工作办法》。黄委印发《黄委水旱灾害防御应急预案（试行）》。淮委编制印发《淮委水旱灾害防御应急预案》。海委编制印发《海河流域水工程防汛抗旱统一调度规定》《海河水利委员会水旱灾害防御应急响应工作规程》。珠江委修订《珠江流域（片）水旱灾害防御应急响应工作规程》。松辽委修订印发《松辽流域主要江河洪水预警发布管理办法（试行）》《松辽委水旱灾害防御应急响应工作规程》。太湖局编制修订《太湖防总应急抢险技术支撑专家库管理办法》《太湖防总防汛抗旱应急响应工作规程》《太湖流域管理局水旱灾害防御应急预案》。

5.2 Rules and Regulations

MWR amended and issued the *Ministry Work Regulations for Flood and Drought Disaster Emergency Response*, and the *Guiding Opinions on Strengthening the Prevention of Flash Floods*. The Changjiang Commission amended and issued the *Emergency Response Plan for Flood and Drought Disaster Prevention in the Yangtze River Basin*, the *Quick Handbook on Flood Control in the Yangtze River*, and the *Commission Measures for the Management of the Working and Expert Groups on Flood and Drought Disaster Prevention*. The Yellow River Commission and the Huaihe Commission both drafted and issued the *Commission Emergency Response Plan for Flood and Drought Disaster Prevention* respectively. The Haihe River Commission compiled and issued the *Unified Dispatch Regulations of Water Projects for Flood and Drought Prevention in the Haihe River Basin* and the *Commission Work Regulations for Flood and Drought Disaster Emergency Response.* The Pearl River Commission revised and issued the *Work Regulations for Flood and Drought Disaster Emergency Response in the Pearl River Basin (Region)*. The Songliao Commission revised and issued the *Management Measures (Trial) for the Major River Flood Warning Release* and the *Commission Work Regulations for Flood and Drought Disaster Emergency Response*. The Taihu Basin Authority drafted and amended the *Management Measures for the Emergency Rescue Technical Expert Database of the Taihu Basin Disaster Control Headquarters*, the *Taihu Work Regulations for Flood and Drought Disaster Emergency Response*, and the *Emergency Response Plan for Flood and Drought Disaster Prevention in the Taihu Lake Basin*.

5.3 方案预案

水利部批复《2022 年长江流域水工程联合调度运用计划》《2022 年雄安新区起步区安全度汛方案》《2022 年夏季引江济太调水方案》《尼尔基水库防洪调度方案》。长江委编制《长江流域水库群抗旱保供水联合调度方案》，完成 51 座控制性水库汛期调度运用计划（方案）的审批和 44 处国家蓄滞洪区运用预案修订。黄委编制修订《洮河应急水量调度预案》《2022 年黄河中下游洪水调度方案》《2022 年黄河上游重要水库群联合防洪调度方案》《2022 年汛前黄河调水调沙预案》，完善滩区人员撤退转移预案和库区、蓄滞洪区运用预案。淮委编制南四湖、骆马湖及新沂河超标洪水防御预案，修订完善淮河流域河道、大型水库汛期调度运用计划。海委编制《海河流域应急水量调度预案》，修订流域各河系超标洪水防御预案。珠江委编制

《2022—2023 年珠江枯水期水量调度实施方案》《2022 年度粤港澳大湾区防洪安全保障方案》。松辽委编制印发《2022 年松花江流域、辽河流域水工程联合调度方案》《2022 年松花江、辽河应急水量调度预案》《察尔森水库防洪调度方案》，批复尼尔基、察尔森、丰满、白山 4 座骨干水库 2022 年汛期调度运用计划。太湖局修订《太湖流域洪水与水量调度方案》。

5.3 Contingency Planning

MWR approved the *2022 Joint Scheduling and Operation Plan for Water Projects in the Yangtze River Basin*, the *2022 Flood Safety Plan for the Start-up Zone of Xiong'an New Area,* the *2022 Water Dispatch Plan for Yangtze-Taihu Water Diversion During Summer*, and the *Flood Control and Scheduling Plan of the Nierji Reservoir*. The Changjiang Commission drafted the *Joint Dispatch Plan of the Reservoir Groups for Drought Prevention and Water Supply in the Yangtze River Basin*; approved the operation plans of 51 controlling reservoirs during flood seasons; and revised the contingency plans for applying 44 national-level flood detention and retention basins. The Yellow River Commission amended and formulated the *Emergency Water Dispatch Plan in the Tao River*, the *2022 Floodwater Regulation Plan for the Middle and Lower Reaches of the Yellow River*, the *2022 Joint Flood Control Plan for Major Reservoirs in the Upstream of the Yellow River*, and the *2022 Pre-flood Flow and Sediment Regulation Plan for Yellow River*. It also modified the contingency plans for the evacuation and relocation of people living in the floodplain areas, and for the application of flood detention and retention basins. The Huaihe Commission drafted the contingency plans for coping with extreme floods in the Nansi Lake, Luoma Lake and Xinyi River. The Commission also improved the flood season operation plans for main waterways and large reservoirs within the basin. The Haihe Commission compiled the *Contingency Plan for Emergency Water Dispatch in the Haihe River Basin*, and revised the contingency plans for preventing extreme flooding in the river systems within the basin. The Pearl River Commission drafted the *2022–2023 Water Dispatch Plan for the Pearl River in Low-Water Period*, and the *2022 Flood Safety Plan for the Greater Bay Area*. The Songliao Commission compiled and issued the *2022 Joint Scheduling Plan for Water Projects in the Songhua River Basin and Liao River Basin*, the *2022 Contingency Plan for Emergency Water Dispatch in the Songhua River and Liao River*, and the *Chaersen Reservoir Floodwater Regulation Plan*. The Commission also approved the 2022 flood season scheduling and operation plans for the four major reservoirs including Nierji, Chaersen, Fengman and Baishan. The Taihu Authority revised the *Dispatch Plan for Flood and Water in the Taihu Basin*.

5.4 信息发布

水利部坚持正面宣传、积极主动发声。部长李国英在全国两会“部长通道”、中央宣传部“中国这十年”系列主题新闻发布会等场合介绍水旱灾害防御相关情况。针对 2021—2022 年度黄河防凌和珠江流域抗旱工作取得全面胜利、水利部两次实施“长江流域水库群抗旱保供水联合调度”专项行动，召开 4 次水利部新闻发布会，集中宣传水旱灾害防御重要举措及成效。编发《汛旱情通报》101 期，及时通过水利部网站、官方微信和水利报社等媒体平台发布汛情旱情实况、预测预报成果、江河洪水和山洪灾害预警、防御工作部署和成效等信息。水利部水旱灾害防御司、信息中心、中国水利水电科学研究院及相关流域管理机构的专家和水利部工作组接受媒体采访 50 余人次，主动回应社会关切，正本清源、解疑释惑。举办“全国水旱灾害防御工作先进典型事迹”“黄河防凌珠江抗旱”“长江中下游抗大旱”3 个图片展览，全景展示水利系统履职尽责、担当使命、坚决打赢水旱灾害防御硬仗的重大行动、先进事迹和感人瞬间。制作发布《洪水来了怎么办》《洪水防御“四个链条”是什么》《关于干旱的那些事》等科普视频和图文，引导公众广泛了解、科学认识洪涝、干旱灾害和相关防御措施，提升全社会防灾减灾意识和能力。围绕 2022 年全国防灾减灾日、国际减灾日，分别以“减轻灾害风险 守护美好家园”“早预警 早行动”为主题，组织开展和参加各项防灾减灾活动，营造水利防灾减灾氛围。

5.4 Information Dissemination

The Ministry adhered to make positive exposure and proactive voice to the public. Minister Li Guoying introduced the efforts put in flood and drought disaster prevention on many occasions such as the "Minister's Passage" at the Two Sessions – the annual sessions of the National People's Congress (NPC) and the Chinese People's Political Consultative Conference (CPPCC) National Committee, and the press conference on "China in the Past Decade"held by CPC Publicity Department. The two rounds of special actions for "joint dispatch of reservoir groups in drought resistance and water supply in the Yangtze River basin" initiated by the Ministry facilitated the full success of 2021–2022 ice flood prevention in the Yellow River and drought control in the Pearl River basin. Four MWR press conferences centrally promoted the key initiatives and effectiveness. MWR also issued 101 volumes of "Flood and Drought Notifications", and published information via multiple media platforms such as MWR website, WeChat official account, *China Water Resources News* about real time conditions of flood and drought, forecasting and prediction results, river flood and flash flood early warning, prevention deployment and performance. Experts from the Department of Flood and Drought Disaster Prevention, Water Resources Information Center, China Institute of Water Resources and Hydropower Research, working groups commissioned by MWR and relevant river basin authorities accepted media interviews for over 50 person-times, actively responding to social concerns, clarifying matters thoroughly, and answering the questions. Three photo exhibitions were held featuring "Advanced and Typical Deeds for National Flood and Drought Disaster Prevention" "Ice Flood Prevention in the Yellow River and Drought Control in the Pearl River", and "Extreme Drought Defense in the Lower and Middle Reaches of the Yellow River", through which the key actions, advanced deeds and memorable moments of the water resources departments were displayed vividly. The Ministry also created and published educative videos and articles with images titled *What to do when flood comes? What are the Four Chains of Flood Prevention? Things You Should Know About Drought* as a way to help the public widely gain knowledge of, and scientifically know about the flood and drought disasters and relevant preventive measures, thus improving the awareness and capacity to disaster prevention and mitigation in the whole society. Echoing with the themes of 2022 National Disaster Prevention and Mitigation Day and the International Day for Disaster Risk Reduction, activities featuring "Reduce disaster risk, protect our beautiful homeland" and "Early warning and early action for all" were organized in a bid to create a participatory environment for disaster prevention and reduction.

5.5 复盘分析

针对2022年极端天气事件多发重发，水利部坚持问题导向、结果导向，全面系统调查分析，客观还原灾害防御过程，深入分析问题原因，总结经验教训，举一反三，及时补短板、堵漏洞、强弱项。开展了2022年西江第4号洪水和北江第2号洪水防御调查评估工作，对珠江流域在防洪工程、非工程体系建设管理方面存在的短板和弱项提出了对策建议；开展了2022年辽河流域洪水防御调查评估工作，对绕阳河曙四联段溃堤、东五家子水库险情等进行了模拟分析与评估，提出了改进措施建议。全面复盘检视了7—8月相继发生的6起典型山洪泥石流灾害事件"四预"工作，深入分析致灾原因，认真查找薄弱环节，研究提出进一步完善山洪灾害防御系统的对策措施，督促指导地方深刻吸取教训。开展了2022年长江流域夏秋连旱防御过程复盘检视工作，针对监测预报、预警、预演、预案，水利工程体系，体制机制等方面存在的问题与差距提出了对策建议。

5.5 Review and Analysis

In recognition of the frequency and severity of extreme weather events in 2022, MWR remained problem-oriented and result-oriented, conducted systematical and thorough investigation and analysis, replayed the real procedures of disaster prevention, tapped into the causes of problems, summarized the lessons learned, finally bolstering weak spots, blocking the loopholes and addressing inadequacies. The Ministry carried out survey and evaluation of the preventive works on the 2022 No.4 flood in the Xijiang River and the 2022 No.2 flood in the Beijiang River, and advised on how to deal with the shortcomings and weaknesses of the flood defense projects, non-engineering system building and management in the Pearl River basin. Survey and evaluation were also conducted in the 2022 flood prevention in the Liaohe River basin. MWR simulated and assessed the dam breach in the Raoyang River and hazards of the Dongwujiazi Reservoir, and proposed enhancements and suggestions accordingly. The Ministry comprehensively reviewed and examined six typical flash floods and flash flood and mud flow events consecutively from July to August in terms of their forecasting, early warning, exercising, and contingency planning by digging into the disaster-inducing factors and the weak links during the process, further proposing countermeasures to improve the flash flood prevention system and reminding the local departments to learn from the lessons. The Ministry also reviewed and examined the preventive process of summer-autumn stretching drought in the Yangtze basin, and suggested on the problems and gaps to be improved related to monitoring, forecasting and early warning, rehearsal and contingency planning, water engineering system, institution and mechanism, etc.

附录
APPENDIX
1950—2022 年
全国水旱灾情统计
STATISTICS OF FLOOD AND DROUGHT
DISASTERS IN CHINA 1950−2022

附表 1 1950—2022 年全国洪涝灾情统计
Appendix 1 Flood disasters and losses 1950−2022

年份 Year	农作物受灾面积 / 千公顷 Affected cropland area /1,000 ha	农作物成灾面积 / 千公顷 Failed cropland area/1,000 ha	因灾死亡人口 / 人 Deaths/person	因灾失踪人口 / 人 Missing persons/ person	倒塌房屋 / 万间 Collapsed dwellings/ 10,000 rooms	直接经济损失 / 亿元 Direct economic loss/100 million RMB
1950	6559.00	4710.00	1982	—	130.50	—
1951	4173.00	1476.00	7819	—	31.80	—
1952	2794.00	1547.00	4162	—	14.50	—
1953	7187.00	3285.00	3308	—	322.00	—
1954	16131.00	11305.00	42447	—	900.90	—
1955	5247.00	3067.00	2718	—	49.20	—
1956	14377.00	10905.00	10676	—	465.90	—
1957	8083.00	6032.00	4415	—	371.20	—
1958	4279.00	1441.00	3642	—	77.10	—
1959	4813.00	1817.00	4540	—	42.10	—
1960	10155.00	4975.00	6033	—	74.70	—
1961	8910.00	5356.00	5074	—	146.30	—
1962	9810.00	6318.00	4350	—	247.70	—
1963	14071.00	10479.00	10441	—	1435.30	—
1964	14933.00	10038.00	4288	—	246.50	—
1965	5587.00	2813.00	1906	—	95.60	—
1966	2508.00	950.00	1901	—	26.80	—
1967	2599.00	1407.00	1095	—	10.80	—
1968	2670.00	1659.00	1159	—	63.00	—
1969	5443.00	3265.00	4667	—	164.60	—
1970	3129.00	1234.00	2444	—	25.20	—
1971	3989.00	1481.00	2323	—	30.20	—
1972	4083.00	1259.00	1910	—	22.80	—
1973	6235.00	2577.00	3413	—	72.30	—

续表 Continued

年份 Year	农作物受灾面积 / 千公顷 Affected cropland area /1,000 ha	农作物成灾面积 / 千公顷 Failed cropland area/1,000 ha	因灾死亡人口 / 人 Deaths/person	因灾失踪人口 / 人 Missing persons/ person	倒塌房屋 / 万间 Collapsed dwellings/ 10,000 rooms	直接经济损失 / 亿元 Direct economic loss/100 million RMB
1974	6431.00	2737.00	1849	—	120.00	—
1975	6817.00	3467.00	29653	—	754.30	—
1976	4197.00	1329.00	1817	—	81.90	—
1977	9095.00	4989.00	3163	—	50.60	—
1978	2820.00	924.00	1796	—	28.00	—
1979	6775.00	2870.00	3446	—	48.80	—
1980	9146.00	5025.00	3705	—	138.30	—
1981	8625.00	3973.00	5832	—	155.10	—
1982	8361.00	4463.00	5323	—	341.50	—
1983	12162.00	5747.00	7238	—	218.90	—
1984	10632.00	5361.00	3941	—	112.10	—
1985	14197.00	8949.00	3578	—	142.00	—
1986	9155.00	5601.00	2761	—	150.90	—
1987	8686.00	4104.00	3749	—	92.10	—
1988	11949.00	6128.00	4094	—	91.00	—
1989	11328.00	5917.00	3270	—	100.10	—
1990	11804.00	5605.00	3589	—	96.60	239.00
1991	24596.00	14614.00	5113	—	497.90	779.08
1992	9423.30	4464.00	3012	—	98.95	412.77
1993	16387.30	8610.40	3499	—	148.91	641.74
1994	18858.90	11489.50	5340	—	349.37	1796.60
1995	14366.70	8000.80	3852	—	245.58	1653.30
1996	20388.10	11823.30	5840	—	547.70	2208.36
1997	13134.80	6514.60	2799	—	101.06	930.11

续表 Continued

年份 Year	农作物受灾面积 / 千公顷 Affected cropland area /1,000 ha	农作物成灾面积 / 千公顷 Failed cropland area/1,000 ha	因灾死亡人口 / 人 Deaths/person	因灾失踪人口 / 人 Missing persons/ person	倒塌房屋 / 万间 Collapsed dwellings/ 10,000 rooms	直接经济损失 / 亿元 Direct economic loss/100 million RMB
1998	22291.80	13785.00	4150	—	685.03	2550.90
1999	9605.20	5389.12	1896	—	160.50	930.23
2000	9045.01	5396.03	1942	—	112.61	711.63
2001	7137.78	4253.39	1605	—	63.49	623.03
2002	12384.21	7439.01	1819	—	146.23	838.00
2003	20365.70	12999.80	1551	—	245.42	1300.51
2004	7781.90	4017.10	1282	—	93.31	713.51
2005	14967.48	8216.68	1660	—	153.29	1662.20
2006	10521.86	5592.42	2276	—	105.82	1332.62
2007	12548.92	5969.02	1230	—	102.97	1123.30
2008	8867.82	4537.58	633	232	44.70	955.44
2009	8748.16	3795.79	538	110	55.59	845.96
2010	17866.69	8727.89	3222	1003	227.10	3745.43
2011	7191.50	3393.02	519	121	69.30	1301.27
2012	11218.09	5871.41	673	159	58.60	2675.32
2013	11777.53	6540.81	775	374	53.36	3155.74
2014	5919.43	2829.99	486	91	25.99	1573.55
2015	6132.08	3053.84	319	81	15.23	1660.75
2016	9443.26	5063.49	686	207	42.77	3643.26
2017	5196.47	2781.19	316	39	13.78	2142.53
2018	6426.98	3131.16	187	32	8.51	1615.47
2019	6680.40	3928.97	573	85	10.30	1922.70
2020	7190.00	4118.21	230	49	9.00	2669.80
2021	4760.43	2643.05	512	78	15.20	2458.92
2022	3413.73	1834.57	143	28	3.13	1288.99

注 2019—2022 年数据来源于应急管理部；“—”表示没有统计数据；因灾失踪人口从 2008 年开始作为指标统计。

Note Data during 2019–2022 are from the Ministry of Emergency Management; “—” means statistics don't exist; missing persons attributed to disasters was determined as a statistical indicator since 2008.

附表 2 2000—2022 年全国中小河流和山洪灾害死亡与失踪人口统计

Appendix 2 Deaths and missing persons attributed to river floods (in small and medium sized rivers) and flash floods 2000−2022

年份 Year	因灾死亡人口 / 人 Deaths/person	因灾失踪人口 / 人 Missing persons/ person	年份 Year	因灾死亡人口 / 人 Deaths/person	因灾失踪人口 / 人 Missing persons/ person
2000	1102	—	2012	473	—
2001	788	—	2013	560	—
2002	924	—	2014	340	—
2003	1307	—	2015	226	50
2004	998	—	2016	481	129
2005	1400	—	2017	207	16
2006	1612	—	2018	129	32
2007	1069	—	2019	347	—
2008	508	—	2020	95	62
2009	430	—	2021	171	67
2010	2824	—	2022	97	22
2011	413	—			

注 “—”表示没有统计数据。

Note “—” means statistics don't exist.

附表 3　1950—2022 年全国干旱灾情统计
Appendix 3　Drought disasters and losses 1950-2022

年份 Year	农作物因旱受灾面积 / 千公顷 Affected cropland area/1,000 ha	农作物因旱成灾面积 / 千公顷 Damaged cropland area/1,000 ha	农作物因旱绝收面积 / 千公顷 Area of crop failure/1,000 ha	因旱粮食损失 / 亿千克 Crop losses/100 million kg	因旱饮水困难人口 / 万人 People with drinking water difficulties/10,000 persons	因旱饮水困难大牲畜 / 万头 Number of bigger-sized livestock having difficulties accessing drinking water/10,000 heads	直接经济损失 / 亿元 Direct economic loss/100 million RMB
1950	2398.00	589.00	—	19.00	—	—	—
1951	7829.00	2299.00	—	36.88	—	—	—
1952	4236.00	2565.00	—	20.21	—	—	—
1953	8616.00	1341.00	—	54.47	—	—	—
1954	2988.00	560.00	—	23.44	—	—	—
1955	13433.00	4024.00	—	30.75	—	—	—
1956	3127.00	2051.00	—	28.60	—	—	—
1957	17205.00	7400.00	—	62.22	—	—	—
1958	22361.00	5031.00	—	51.28	—	—	—
1959	33807.00	11173.00	—	108.05	—	—	—
1960	38125.00	16177.00	—	112.79	—	—	—
1961	37847.00	18654.00	—	132.29	—	—	—
1962	20808.00	8691.00	—	89.43	—	—	—
1963	16865.00	9021.00	—	96.67	—	—	—
1964	4219.00	1423.00	—	43.78	—	—	—
1965	13631.00	8107.00	—	64.65	—	—	—
1966	20015.00	8106.00	—	112.15	—	—	—
1967	6764.00	3065.00	—	31.83	—	—	—
1968	13294.00	7929.00	—	93.92	—	—	—
1969	7624.00	3442.00	—	47.25	—	—	—
1970	5723.00	1931.00	—	41.50	—	—	—
1971	25049.00	5319.00	—	58.12	—	—	—
1972	30699.00	13605.00	—	136.73	—	—	—
1973	27202.00	3928.00	—	60.84	—	—	—

续表 Continued

年份 Year	农作物因旱受灾面积 / 千公顷 Affected cropland area/1,000 ha	农作物因旱成灾面积 / 千公顷 Damaged cropland area/1,000 ha	农作物因旱绝收面积 / 千公顷 Area of crop failure/1,000 ha	因旱粮食损失 / 亿千克 Crop losses/ 100 million kg	因旱饮水困难人口 / 万人 People with drinking water difficulties/ 10,000 persons	因旱饮水困难大牲畜 / 万头 Number of bigger-sized livestock having difficulties accessing drinking water/10,000 heads	直接经济损失 / 亿元 Direct economic loss/100 million RMB
1974	25553.00	2296.00	—	43.23	—	—	—
1975	24832.00	5318.00	—	42.33	—	—	—
1976	27492.00	7849.00	—	85.75	—	—	—
1977	29852.00	7005.00	—	117.34	—	—	—
1978	40169.00	17969.00	—	200.46	—	—	—
1979	24646.00	9316.00	—	138.59	—	—	—
1980	26111.00	12485.00	—	145.39	—	—	—
1981	25693.00	12134.00	—	185.45	—	—	—
1982	20697.00	9972.00	—	198.45	—	—	—
1983	16089.00	7586.00	—	102.71	—	—	—
1984	15819.00	7015.00	—	106.61	—	—	—
1985	22989.00	10063.00	—	124.04	—	—	—
1986	31042.00	14765.00	—	254.34	—	—	—
1987	24920.00	13033.00	—	209.55	—	—	—
1988	32904.00	15303.00	—	311.69	—	—	—
1989	29358.00	15262.00	2423.33	283.62	—	—	—
1990	18174.67	7805.33	1503.33	128.17	—	—	—
1991	24914.00	10558.67	2108.67	118.00	4359.00	6252.00	—
1992	32980.00	17048.67	2549.33	209.72	7294.00	3515.00	—
1993	21098.00	8658.67	1672.67	111.80	3501.00	1981.00	—
1994	30282.00	17048.67	2526.00	233.60	5026.00	6012.00	—
1995	23455.33	10374.00	2121.33	230.00	1800.00	1360.00	—
1996	20150.67	6247.33	686.67	98.00	1227.00	1675.00	—
1997	33514.00	20010.00	3958.00	476.00	1680.00	850.00	—
1998	14237.33	5068.00	949.33	127.00	1050.00	850.00	—
1999	30153.33	16614.00	3925.33	333.00	1920.00	1450.00	—

续表 Continued

年份 Year	农作物因旱受灾面积 / 千公顷 Affected cropland area/1,000 ha	农作物因旱成灾面积 / 千公顷 Damaged cropland area/1,000 ha	农作物因旱绝收面积 / 千公顷 Area of crop failure/1,000 ha	因旱粮食损失 / 亿千克 Crop losses/ 100 million kg	因旱饮水困难人口 / 万人 People with drinking water difficulties/ 10,000 persons	因旱饮水困难大牲畜 / 万头 Number of bigger-sized livestock having difficulties accessing drinking water/10,000 heads	直接经济损失 / 亿元 Direct economic loss/100 million RMB
2000	40540.67	26783.33	8006.00	599.60	2770.00	1700.00	—
2001	38480.00	23702.00	6420.00	548.00	3300.00	2200.00	—
2002	22207.33	13247.33	2568.00	313.00	1918.00	1324.00	—
2003	24852.00	14470.00	2980.00	308.00	2441.00	1384.00	—
2004	17255.33	7950.67	1677.33	231.00	2340.00	1320.00	—
2005	16028.00	8479.33	1888.67	193.00	2313.00	1976.00	—
2006	20738.00	13411.33	2295.33	416.50	3578.23	2936.25	986.00
2007	29386.00	16170.00	3190.67	373.60	2756.00	2060.00	1093.70
2008	12136.80	6797.52	811.80	160.55	1145.70	699.00	545.70
2009	29258.80	13197.10	3268.80	348.49	1750.60	1099.40	1206.59
2010	13258.61	8986.47	2672.26	168.48	3334.52	2440.83	1509.18
2011	16304.20	6598.60	1505.40	232.07	2895.45	1616.92	1028.00
2012	9333.33	3508.53	373.80	116.12	1637.08	847.63	533.00
2013	11219.93	6971.17	1504.73	206.36	2240.54	1179.35	1274.51
2014	12271.70	5677.10	1484.70	200.65	1783.42	883.29	909.76
2015	10067.05	5577.04	1005.39	144.41	836.43	806.77	579.22
2016	9872.76	6130.85	1018.20	190.64	469.25	649.73	484.15
2017	9946.43	4490.02	752.71	134.44	477.78	514.29	437.88
2018	7397.21	3667.23	610.21	156.97	306.69	462.30	483.62
2019	7838.00	4760.17	1113.60	92.29	692.29	368.10	457.40
2020	5081.00	2759.08	704.50	123.04	668.98	448.63	249.20
2021	3426.16	1949.00	464.12	49.28	546.35	250.63	200.87
2022	6090.21	2858.39	611.78	57.44	542.38	331.92	512.85

注 2019 — 2022 年数据第 2、3、4、8 列来源于应急管理部；第 2 列“农作物因旱受灾面积”2019 年之前为“作物因旱受灾面积”；第 3 列“农作物因旱成灾面积”2019 年之前为“作物因旱成灾面积”；第 4 列“农作物因旱绝收面积”2019 年之前为“作物因旱绝收面积”；“—”表示没有统计数据。

Note Data during 2019–2022 in columns 2, 3, 4 & 8 are from the Ministry of Emergency Management; “—” means statistics don't exist.